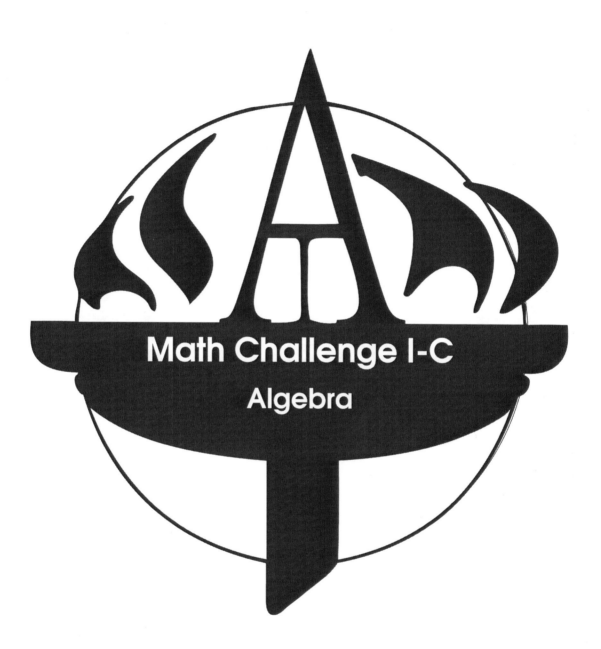

Math Challenge I-C
Algebra

Areteem Institute

Math Challenge I-C Algebra

Edited by   John Lensmire
              David Reynoso
              Kelly Ren
              Kevin Wang, Ph.D.

ISBN: 1-944863-23-0
ISBN-13: 978-1-944863-23-4

First printing, March 2019.

TITLES PUBLISHED BY ARETEEM PRESS

Cracking the High School Math Competitions (and Solutions Manual) - Covering AMC 10 & 12, ARML, and ZIML
Mathematical Wisdom in Everyday Life (and Solutions Manual) - From Common Core to Math Competitions
Geometry Problem Solving for Middle School (and Solutions Manual) - From Common Core to Math Competitions
Fun Math Problem Solving For Elementary School (and Solutions Manual)

ZIML MATH COMPETITION BOOK SERIES

ZIML Math Competition Book Division E 2016-2017
ZIML Math Competition Book Division M 2016-2017
ZIML Math Competition Book Division H 2016-2017
ZIML Math Competition Book Jr Varsity 2016-2017
ZIML Math Competition Book Varsity Division 2016-2017
ZIML Math Competition Book Division E 2017-2018
ZIML Math Competition Book Division M 2017-2018
ZIML Math Competition Book Division H 2017-2018
ZIML Math Competition Book Jr Varsity 2017-2018
ZIML Math Competition Book Varsity Division 2017-2018

MATH CHALLENGE CURRICULUM TEXTBOOKS SERIES

Math Challenge I-A Pre-Algebra and Word Problems
Math Challenge I-B Pre-Algebra and Word Problems
Math Challenge I-C Algebra
Math Challenge II-A Algebra
Math Challenge II-B Algebra
Math Challenge III Algebra
Math Challenge I-A Geometry
Math Challenge I-B Geometry
Math Challenge I-C Topics in Algebra
Math Challenge II-A Geometry
Math Challenge II-B Geometry
Math Challenge III Geometry
Math Challenge I-A Counting and Probability
Math Challenge I-B Counting and Probability
Math Challenge I-C Geometry

Math Challenge II-A Combinatorics
Math Challenge II-B Combinatorics
Math Challenge III Combinatorics
Math Challenge I-A Number Theory
Math Challenge I-B Number Theory
Math Challenge I-C Finite Math
Math Challenge II-A Number Theory
Math Challenge II-B Number Theory
Math Challenge III Number Theory

COMING SOON FROM ARETEEM PRESS

Fun Math Problem Solving For Elementary School Vol. 2 (and Solutions Manual)
Counting & Probability for Middle School (and Solutions Manual) - From Common Core to Math Competitions
Number Theory Problem Solving for Middle School (and Solutions Manual) - From Common Core to Math Competitions

The books are available in paperback and eBook formats (including Kindle and other formats).
To order the books, visit https://areteem.org/bookstore.

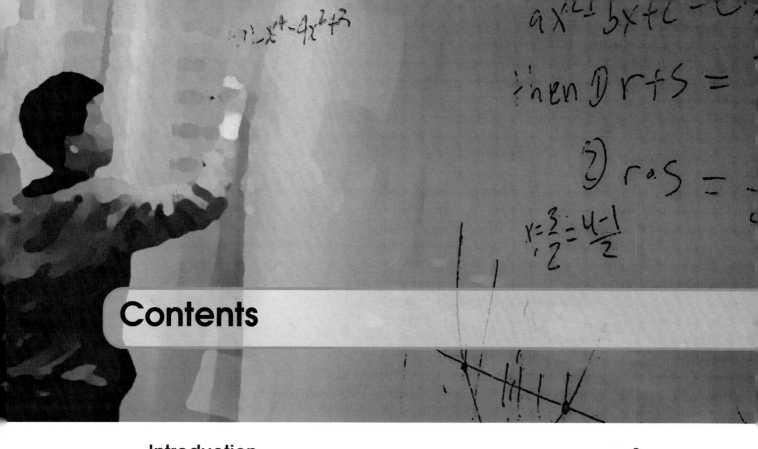

Contents

Introduction

Math Challenge I-C is a four-part course designed to bridge the middle school and high school math materials. For students who participate in the American Math Competitions (AMC), there is a big gap in both the fundamental math concepts and the problem-solving techniques involved between the AMC 8 and AMC 10 contests. This course is developed to help students transition smoothly from middle school to high school, and prepare them for high school math competitions including the AMC 10 & 12, ARML, and ZIML. The full course covers topics and introductory problem solving in algebra, geometry, and finite math. Algebraic topics include linear equations, systems of equations and inequalities, exponents and radicals, factoring polynomials, and solving quadratic equations. Geometric topics include angles in triangles, quadrilaterals, and polygons, congruent and similar polygons, calculating area, and algebraic geometry. Topics in finite math include logic, introductory number theory, and an introduction to probability and statistics. These topics serve as the fundamental knowledge needed for a more advanced problem solving course such as Math Challenge II-A.

The course is divided into four terms:

- Summer, covering Algebra
- Fall, covering covering additional topics in Algebra
- Winter, covering Geometry
- Spring, covering Finite Math

The book contains course materials for Math Challenge I-C: Algebra.

We recommend that students take all four terms starting with the Summer, but students with the required background are welcome to join for later terms in the course.

Students can sign up for the online live or self-paced course at `classes.areteem.org`.

About Areteem Institute

Areteem Institute is an educational institution that develops and provides in-depth and advanced math and science programs for K-12 (Elementary School, Middle School, and High School) students and teachers. Areteem programs are accredited supplementary programs by the Western Association of Schools and Colleges (WASC). Students may attend the Areteem Institute in one or more of the following options:

- Live and real-time face-to-face online classes with audio, video, interactive online whiteboard, and text chatting capabilities;
- Self-paced classes by watching the recordings of the live classes;
- Short video courses for trending math, science, technology, engineering, English, and social studies topics;
- Summer Intensive Camps held on prestigious university campuses and Winter Boot Camps;
- Practice with selected free daily problems and monthly ZIML competitions at ziml.areteem.org.

Areteem courses are designed and developed by educational experts and industry professionals to bring real world applications into STEM education. The programs are ideal for students who wish to build their mathematical strength in order to excel academically and eventually win in Math Competitions (AMC, AIME, USAMO, IMO, ARML, MathCounts, Math Olympiad, ZIML, and other math leagues and tournaments, etc.), Science Fairs (County Science Fairs, State Science Fairs, national programs like Intel Science and Engineering Fair, etc.) and Science Olympiads, or for students who purely want to enrich their academic lives by taking more challenging courses and developing outstanding analytical, logical, and creative problem solving skills.

Since 2004 Areteem Institute has been teaching with methodology that is highly promoted by the new Common Core State Standards: stressing the conceptual level understanding of the math concepts, problem solving techniques, and solving problems with real world applications. With the guidance from experienced and passionate professors, students are motivated to explore concepts deeper by identifying an interesting problem, researching it, analyzing it, and using a critical thinking approach to come up with multiple solutions.

Thousands of math students who have been trained at Areteem have achieved top honors and earned top awards in major national and international math competitions, including Gold Medalists in the International Math Olympiad (IMO), top winners and qualifiers at the USA Math Olympiad (USAMO/JMO) and AIME, top winners at the

Zoom International Math League (ZIML), and top winners at the MathCounts National Competition. Many Areteem Alumni have graduated from high school and gone on to enter their dream colleges such as MIT, Cal Tech, Harvard, Stanford, Yale, Princeton, U Penn, Harvey Mudd College, UC Berkeley, or UCLA. Those who have graduated from colleges are now playing important roles in their fields of endeavor.

Further information about Areteem Institute, as well as updates and errata of this book, can be found online at http://www.areteem.org.

About Zoom International Math League

The Zoom International Math League (ZIML) has a simple goal: provide a platform for students to build and share their passion for math and other STEM fields with students from around the globe. Started in 2008 as the Southern California Mathematical Olympiad, ZIML has a rich history of past participants who have advanced to top tier colleges and prestigious math competitions, including American Math Competitions, MATHCOUNTS, and the International Math Olympaid.

The ZIML Core Online Programs, most available with a free account at `ziml.areteem.org`, include:

- **Daily Magic Spells:** Provides a problem a day (Monday through Friday) for students to practice, with full solutions available the next day.
- **Weekly Brain Potions:** Provides one problem per week posted in the online discussion forum at `ziml.areteem.org`. Usually the problem does not have a simple answer, and students can join the discussion to share their thoughts regarding the scenarios described in the problem, explore the math concepts behind the problem, give solutions, and also ask further questions.
- **Monthly Contests:** The ZIML Monthly Contests are held the first weekend of each month during the school year (October through June). Students can compete in one of 5 divisions to test their knowledge and determine their strengths and weaknesses, with winners announced after the competition.
- **Math Competition Practice:** The Practice page contains sample ZIML contests and an archive of AMC-series tests for online practice. The practices simulate the real contest environment with time-limits of the contests automatically controlled by the server.
- **Online Discussion Forum:** The Online Discussion Forum is open for any comments and questions. Other discussions, such as hard Daily Magic Spells or the Weekly Brain Potions are also posted here.

These programs encourage students to participate consistently, so they can track their progress and improvement each year.

In addition to the online programs, ZIML also hosts onsite Local Tournaments and Workshops in various locations in the United States. Each summer, there are onsite ZIML Competitions at held at Areteem Summer Programs, including the National ZIML Convention, which is a two day convention with one day of workshops and one day of competition.

ZIML Monthly Contests are organized into five divisions ranging from upper elementary school to advanced material based on high school math.

- **Varsity:** This is the top division. It covers material on the level of the last 10 questions on the AMC 12 and AIME level. This division is open to all age levels.
- **Junior Varsity:** This is the second highest competition division. It covers material at the AMC 10/12 level and State and National MathCounts level. This division is open to all age levels.
- **Division H:** This division focuses on material from a standard high school curriculum. It covers topics up to and including pre-calculus. This division will serve as excellent practice for students preparing for the math portions of the SAT or ACT. This division is open to all age levels.
- **Division M:** This division focuses on problem solving using math concepts from a standard middle school math curriculum. It covers material at the level of AMC 8 and School or Chapter MathCounts. This division is open to all students who have not started grade 9.
- **Division E:** This division focuses on advanced problem solving with mathematical concepts from upper elementary school. It covers material at a level comparable to MOEMS Division E. This division is open to all students who have not started grade 6.

To participate in the ZIML Online Programs, create a free account at `ziml.areteem.org`. The ZIML site features are also provided on the ZIML Mobile App, which is available for download from Apple's App Store and Google Play Store.

Acknowledgments

This book contains many years of collaborative work by the staff of Areteem Institute. This book could not have existed without their efforts. Huge thanks go to the Areteem staff for their contributions!

The examples and problems in this book were either created by the Areteem staff or adapted from various sources, including other books and online resources. Especially, some good problems from previous math competitions and contests such as AMC, AIME, ARML, MATHCOUNTS, and ZIML are chosen as examples to illustrate concepts or problem-solving techniques. The original resources are credited whenever possible. However, it is not practical to list all such resources. We extend our gratitude to the original authors of all these resources.

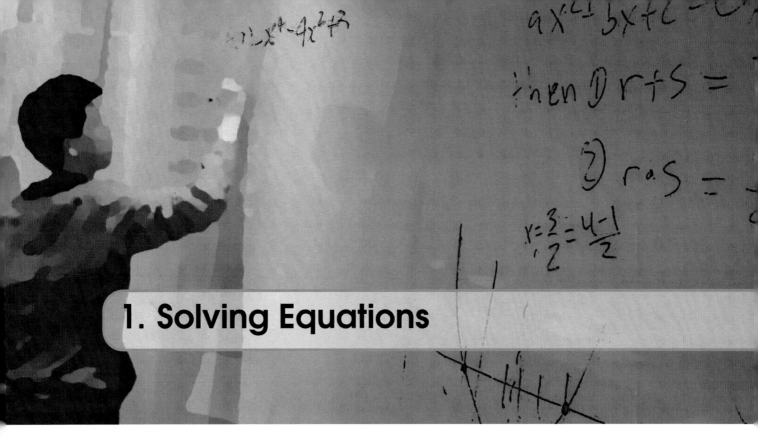

1. Solving Equations

Basic Definitions and Properties

- **Associative Property**
  - For Addition: $(a+b)+c = a+(b+c)$.
  - For Multiplication: $(a \times b) \times c = a \times (b \times c)$.
- **Commutative Property**
  - For Addition: $a+b = b+a$.
  - For Multiplication: $a \times b = b \times a$.
- **Distributive Property**
  - $a \times (b+c) = a \times b + a \times c$
  - Also in the form (because multiplication is commutative) $(a+b) \times c = a \times c + b \times c$.
- **Numbers and Variables**
  - There are various types of numbers we will use through this course: integers, fractions, decimals, real numbers, etc.
  - We can represent unknown numbers using 'variables' (such as x, y, a, p, etc.)
  - We can combine the same variables (often called 'like terms'). For example, $3x + 4x = 7x$. Note that $3x + 4y$ cannot be combined.

Expressions and Equations

- **Expressions**
    - ○ The result of combining numbers and variables using operations (such as $+, -, \times, \div$) is called an 'expression'.
    - ○ For example, $2 + 3$, $x + y$, $\dfrac{x+4}{5}$ are all expressions.
- **Equations**
    - ○ Setting two expressions equal to each other gives an 'equation'.
    - ○ For example, $4 + 5 = 5 + 4$, $3x = 15$, and $3x + y = 15$ are all equations.
    - ○ The main goal of today will be to solve simple equations. For example, we call $x = 5$ a solution to the equation $3x = 15$ because $3 \times 5 = 15$.
- **Properties of Equality**
    - ○ **Addition** If $a = b$ then $a + c = b + c$. That is, given an equation, adding the same number to both sides does not change the solution to the equation.
    - ○ **Multiplication** If $a = b$ and $c \neq 0$ then $a \times c = b \times c$. That is, given an equation, multiplying both sides by a non-zero number does not change the solution to the equation.
- **Inverses**
    - ○ **Additive** Every number has an 'additive inverse'. The additive inverse of 3 is -3 and the additive inverse of -1.5 is 1.5 because $3 + (-3) = (-1.5) + 1.5 = 0$. In general, the additive inverse of a is $-a$.
    - ○ **Multiplicative** Every non-zero number has an 'multiplicative inverse'. The multiplicative inverse of 3 is $\dfrac{1}{3}$ and the multiplicative inverse of $\dfrac{3}{2}$ is $\dfrac{2}{3}$ because $3 \times \dfrac{1}{3} = \dfrac{3}{2} \times \dfrac{2}{3} = 1$. In general, the multiplicative inverse of a is $\frac{1}{a}$.
    - ○ Inverses are very useful in solving equations, because they allow us to isolate (or solve for) the variables.

1.1   Example Questions

Problem 1.1 Evaluate the following expressions.

(a) $3 \times (2 + x)$ if $x = 4$.

(b) $(k-1)(k+2)$ if $k=-1$.

(c) $p^2+\dfrac{9}{p}$ if $p=3$.

(d) $(a-b)(a+b)$ if $a=5$ and $b=-3$.

(e) $\dfrac{2x+3}{y-x}$ if $x=2$ and $y=6$.

Problem 1.2 For each of the equations below, determine whether the given variables are solutions to the equation.

(a) $-\dfrac{1}{2}x+4=2$ if $x=4$.

(b) $3x-7=2x+1$ if $x=5$.

(c) $3(x-2)=2(x+2)$ if $x=10$.

(d) $\dfrac{x+5}{x-5}=0$ if $x=5$.

Problem 1.3 Simplify the following, combining like terms where possible.

(a) $5m - 4 - (m + 3)$

(b) $0.2(2x + 4) - 0.05(5x - 4)$

(c) $4 - (6 + 3r - 2s)$

(d) $2(x + y) + 3(2 - y) + 2$

Problem 1.4 Solve the following equations.

(a) $x - 5 = 10$.

(b) $y + 3 = 1$.

(c) $3t = 78$

(d) $4 = \dfrac{z}{6}$

Problem 1.5 Solve the following.

(a) $4 - h = 12$.

(b) $3(x-3) = 6$

Problem 1.6 Solve the following equations.

(a) $\dfrac{4x}{5} = 20.$

(b) $3(c-0.78) - 2c = 3.57.$

(c) $3(2s-3) = 4s + 3 - (-2s+1)$

(d) Solve $\dfrac{2}{3}a - 5 = \dfrac{1}{3}a + 2.$

Problem 1.7 Solve the following equations for the specified variable.

(a) $3x - 4y = 12$ for x.

(b) $x(y+2) = 3$ for y.

(c) $\dfrac{2}{p} = \dfrac{q}{3}$ for p.

Problem 1.8 For each of the following word problems, write and equation and then solve that equation to solve the problem.

(a) George hires a gardener for his house. The gardener charges an initial fee of $20 plus $15 per hour that he works. If George pays a total of $95 to the gardener, how many hours does the gardener work?

(b) A rectangle is 3 inches longer than it is wide. If the rectangle has a perimeter of 26 inches, what is the area of the rectangle in square inches?

Problem 1.9 Solve the following motion problems. Recall that distance can be calculated using the equation $d = r \times t$ (distance equals rate times time).

(a) William drove to visit his parents. At the start of his trip, there was no traffic, and he drove for 4 hours at his normal speed. For the last 2 hours of the trip, traffic caused William to drive 20 mph less than his normal speed. If William drove 350 miles in total, what is his normal driving speed in mph?

(b) Parker and Candace run a race. Parker cheats and starts 2 seconds before Candace, but they still end up tying during the race. If Parker can run 6 m/s and Candace can run 6.5 m/s, how long was the race in meters?

Problem 1.10 Solve the following problems related to percents.

(a) Laura and Ivy went shopping for clothes during their huge annual sale. Before the discounts, Laura's total was $100 and Ivy's was $80. After the discounts, they remarked that Laura saved $10 more than Ivy. What percentage discount was the store offering during the sale?

(b) Natasha invested $10,000, split between stocks and bonds. After one year, the stocks had an annual yield of 8% and the bonds had an annual yield of 2%. If Natasha made $560 on the stocks during the year, how many dollars did she invest in bonds?

1.2 Quick Response Questions

Problem 1.11 Which of the following properties justifies

$$2 \times (3+6) = 6+12?$$

(A) Associative Property For Addition
(B) Commutative Property For Addition
(C) Distributive Property
(D) Associative Property For Multiplication

Problem 1.12 Which of the following properties justifies

$$5 \times (2x) = (10)x?$$

(A) Distributive Property
(B) Associative Property For Multiplication
(C) Commutative Property For Multiplication
(D) Associative Property For Addition

Problem 1.13 Which of the following properties justifies

$$10x + 5 = 5 \times (1+2x)?$$

(A) Distributive Property and Associative Property for Addition
(B) Distributive Property and Commutative Property for Addition
(C) Distributive Property and Commutative Property for Multiplication
(D) Associative Properties for Addition and Multiplication

Problem 1.14 What is the additive inverse of 4?

Problem 1.15 What is the multiplicative inverse of 0.25?

Problem 1.16 If we simplify and combine like terms

$$5(x+2y)+3 =$$

(A) 10xy+3
(B) 5x+10y+3
(C) 10xy+15
(D) 5x+10y+15

Problem 1.17 If we simplify and combine like terms

$$2(q+1)-4(3-q) =$$

(A) q-11
(B) -2q-10
(C) 6q-10
(D) 6q+14

Problem 1.18 Evaluate $\dfrac{a+2b}{4-b}$ if $a = -1$ and $b = 3$. Round your answer to the nearest hundredth if necessary.

Problem 1.19 Solve the equation $t - 1.75 = 2$. What is t?

Problem 1.20 Solve the equation $-3p = 27$. What is p?

1.3 Practice Questions

Problem 1.21 Evaluate the following expresions if $x = 3$, $y = 2$, and $z = -1$.

(a) $3(x+2)+4y$

(b) $\dfrac{x-5}{2}+y-z$

Problem 1.22 Answer the following.

(a) Is $k = -2$ a solution to $2(k+2) = 8+4k$?

(b) Is $w = 4$ a solution to $\dfrac{w}{2} = \dfrac{8}{w}$?

Problem 1.23 Combine like terms to simplify the following

(a) $(a+b+4)+(2a+3b+5)$.

(b) $\dfrac{1}{2}(s-4)-\dfrac{1}{4}(s-t)$.

Problem 1.24 Solve the equation $3s-4 = 14$.

Problem 1.25 Solve the equation $-2(4-x) = 8$.

Problem 1.26 Solve the equation $\frac{2}{3}(p-6) = \frac{1}{3}p + 2$.

Problem 1.27 Solve $2w + 4 = 3w + t$ for w.

Problem 1.28 Recall that the three angles in a triangle add up to $180°$. A triangle with angles $\angle A$, $\angle B$, and $\angle C$ has a smallest angle of $\angle A$. Suppose $\angle B$ is twice that of $\angle A$ and $\angle C$ is $40°$ more than $\angle A$. What is the measure of $\angle A$ in degrees?

Problem 1.29 Terry drove from Los Angeles to San Francisco, averaging 55 miles per hour. When he drove back to Los Anglees from San Francisco (using the same route), the journey took 3 hours longer because he only averages 40 miles per hour. How long did his initial trip from Los Angeles to San Francisco take in hours?

Problem 1.30 Larry went shopping over Memorial Day Weekend and, before any discounts, had a total bill of $500. Some of his items were electronics, which were 40% off for the weekend. The rest of the items were 20% off for the weekend. If Larry saved $150 in total, what was the original price (before discounts) of all the electronics he purchased?

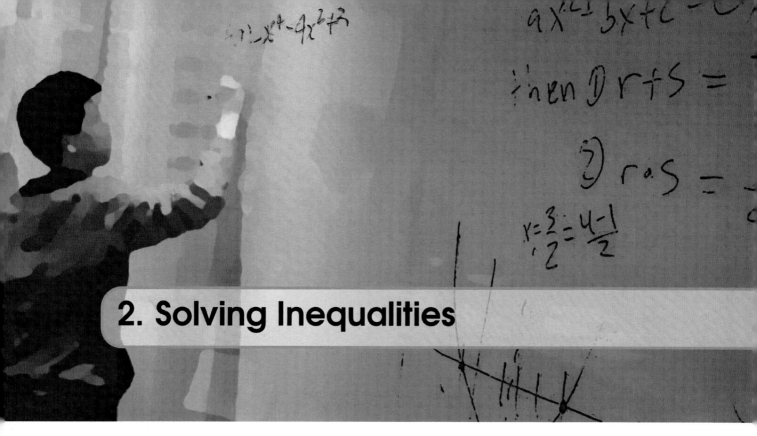

2. Solving Inequalities

Inequalities

- Replacing $=$ (equals) with $<, \leq$ (less than, less than or equal to) or $>, \geq$ (greater than, greater than or equal to) in an equation gives an inequality.
- Solving inequalities is very similar to solving equations, but we need to be careful with negative numbers. For example,
  - $x + 4 > 2$ is the same as $x > -2$.
  - $x - 2 > 4$ is the same as $x > 6$.
  - $\frac{2}{3}x > 1$ is the same as $x > \frac{3}{2}$.
  - $-2x > 4$ is the same as $x < -2$. Note: Here when multiplying or dividing by a negative number we need to 'flip' the inequality.

The Number Line and Interval Notation

- It is often useful to graph solutions to equations or inequalities using a number line.
- A number line contains all real numbers: whole numbers, fractions and decimals, and also irrational numbers. Recall irrational numbers are those that cannot be written as a fraction, such as π or $-\sqrt{3}$.
- We use 'interval notation' to help write inequalities in a short way. Parentheses are used for 'open' intervals (strict inequalities $<$ or $>$) while square brackets are used for 'closed' intervals ($\leq$ or $\geq$). For example,
  - $2 < x$ and $x < 3$, which is the same as $2 < x < 3$, can be written as the interval $(2, 3)$.

- ○ $-4 \leq x \leq 2$ can be written as the interval $[-4, 2]$.
- Using ∞ or $-\infty$ we can also have intervals such as $x \leq 2$ written as $(-\infty, 2]$ or $x > 3$ written as $(3, \infty)$. Note we always use the parentheses (or) for $\pm\infty$ because infinity is not a real number.

Compound Inequalities

- Multiple inequalities combined with 'and' or 'or' are called 'compound inequalities'. We've already seen one example, where $2 < x$ and $x < 3$ could be written as $2 < x < 3$ or the interval $(2, 3)$.
- We use the symbol $\cap$ to denote the 'intersection' of two intervals. A number is in the intersection of two intervals if it is in both intervals.
  - ○ The example we've already seen, $2 < x$ and $x < 3$ can be written as the intersection $(2, \infty) \cap (-\infty, 3) = (2, 3)$.
  - ○ The intersection $[2, 10] \cap (5, \infty) = (5, 10]$ can be thought of as the collection of x such that $2 \leq x \leq 10$ and $x > 5$ which is the same as $5 < x \leq 10$.
  - ○ Note that intersection is often used with the word 'and'.
- We use the symbol $\cup$ to denote the 'union' of two intervals. A number is in the union of two intervals if it is in one of the two intervals.
  - ○ The collection of x such that $x < -2$ or $x > 2$ can be written as the union $(-\infty, -2) \cup (2, \infty)$.
  - ○ The union $[2, 10] \cup (5, \infty) = [2, \infty)$ can be thought of as the collection of x such that $2 \leq x \leq 10$ or $x > 5$ which is the same as $x \geq 2$.
  - ○ Note that intersection is often used with the word 'or'.

2.1   Example Questions

Problem 2.1 Inequality Warmups: True or False

(a) $-3 \leq 3 < -2$.

(b) The inequalities are $-3 < x$ and $x > -3$ are the same.

(c) The inequalities $x + 2 > 3$ is the same as $2x + 4 > 8$.

(d) The inequalities $-x > -2$ and $x < 2$ are the same.

Problem 2.2 Write each of the following inequalities in interval notation and sketch its graph.

(a) $x \geq 3$

(b) $x < 10$ and $x > -5$.

(c) $x < 2$ or $x > 8$

(d) $x < -5$ or $x < 10$

Problem 2.3 For each of the inequalities below, determine whether the number satisfies the inequality.

(a) $4x - 5 < 7$ if $x = 2$.

(b) $-4.25x + 32.58 < -1.73$ if $x = 8$.

(c) $-10x + 9 \leq 4(x + 2)$ if $x = 1.5$

Problem 2.4 Solve the following inequalities for the given variable.

(a) $2x - y > 4$ for y.

(b) $\dfrac{2}{3}a - 8b \le \dfrac{1}{4}$ for a.

Problem 2.5 Humans can roughly hear sounds in the range of $31 - 19{,}000$ hertz. Cats on the other hand can hear sounds in the range $55 - 77{,}000$ hertz.

(a) Write the range (in hertz) humans can hear in interval notation. Write the range (in hertz) cats can hear in interval notation.

(b) Write the range (in hertz) of sounds that can be heard by both humans and cats in interval notation.

(c) Write the range (in hertz) of sounds that can be heard by either humans or cats in interval notation.

Problem 2.6 For each of the following, write and solve an inequality to help answer the question.

(a) A soccer field is 100% longer than it is wide. If the perimeter of the soccer field must be at least 180 meters, what is the range of possible widths for the field?

(b) Paul and Peter want to buy their father an autographed poster for his birthday. They are looking at posters that range between $80 and $120. Paul is the older brother, and offers to pay $20 more than Peter for the gift. What is the range of money Peter will have to pay for the gift?

(c) Recall that temperatures can be measured in Fahrenheit or in Celsius, where $F = \frac{9}{5}C + 32$ is a formula relating the temperature in Fahrenheit (F) to the temperature in Celsius (C). George does not like the heat, and will not play outside if the temperature is above 95 degrees Fahrenheit. What range of temperatures is too hot for George to play outside measured in Celsius?

Problem 2.7 Solve the following

(a) Harry's grade in his history class is based on the average of 4 tests. On the first three tests he got scores of 45, 75, and 80. Harry wants to get at least a C in the class, so his final grade needs to be at least 70. What is the range of scores Harry can get on the fourth test to ensure he gets at least a C?

(b) David and his friends are hungry and order a large pizza for delivery. A pizza costs $11 and each additional topping is $1.50. With tax and tip for the delivery, the final cost is 25% more than the price of the actual pizza. David has at most $20 to spend on the pizza. How many toppings can he and his friends get?

Problem 2.8 How many different sets of four consecutive positive even integers whose sum is less than 100 are there in total?

Problem 2.9 Solve the inequality $x^2 + 4 \geq 8$.

Problem 2.10 Find all z such that $10 - z^2 > -71$ and $5z - 3 > 2(z + 2)$.

2.2 Quick Response Questions

Problem 2.11 Is it true that $-3 \times 4 - 1 < 1 - 5$?

Problem 2.12 Which of the following fills in the blank so that the two inequalities below have the same solutions?

$$2x + 3 < 0 \text{ equivalent to } \frac{-3}{2} \underline{\quad} x$$

(A) $<$
(B) $\leq$
(C) $>$
(D) $\geq$

Problem 2.13 Which of the following fills in the blank so that the two inequalities below have the same solutions?

$$-3x > 18 \text{ equivalent to } x \underline{\quad} 6$$

(A) $<$
(B) $\leq$
(C) $>$
(D) $\geq$

Problem 2.14 Which of the following intervals is the same as $x < 7$ and $x < 13$?

(A) $(-\infty, 13)$
(B) $(-\infty, 7)$
(C) $(7, 13)$
(D) $(-\infty, 7) \cup (13, \infty)$

Problem 2.15 Is $x = 0$ in the interval $(-\infty, 5) \cap [-2, 10]$?

Problem 2.16 The interval $(-\infty, 5) \cup [-2, 10]$ can be written as $x \leq K$ for some integer K. What is K?

Problem 2.17 The solutions to the inequality $5 - 9p > 9 - 8p$ can be written as $p < L$ for some number L. What is L, rounded to the nearest tenth if necessary?

Problem 2.18 Which of the following describes the solutions to $-2 \leq 5 - k \leq 9$?

(A) $-7 \leq k \leq 4$
(B) $7 \leq k \leq -4$
(C) $k \leq 4$ or $k \geq 7$
(D) $-4 \leq k \leq 7$

Problem 2.19 Alice and Grace go shopping. Alice spends $20 more dollars than Grace. Together the pair spends less than $100. If g denotes the dollars spent by Grace, which inequality best describes the information given?

(A) $g + g + 20 \leq 100$
(B) $g + g + 20 < 100$
(C) $g + g - 20 < 100$
(D) $g + g + 20 > 100$

Problem 2.20 Which of the following describes the solutions to $3x^2 < 75$?

(A) $-5 < x$ or $x > 5$
(B) $x < -5$ and $x < 5$
(C) $-5 < x$ or $x < 5$
(D) $-5 < x$ and $x < 5$

2.3 Practice Questions

Problem 2.21 Solve the following inequalities

(a) $3x + 9 \le 3(x + 4)$

(b) $9 - x > 9 + x.$

Problem 2.22 Write the following in interval notation and sketch its graph.

(a) $x < 6$ and $x \ge -3$.

(b) $x \ge 4$ and $x^2 > 9$.

Problem 2.23 Solve the inequality $2x + 1.5 > -1.4(x - 3)$.

Problem 2.24 Solve the inequality $3(a + b) \ge 4(5 - a)$ for b.

Problem 2.25 Sarah and her friend Laura go out to dinner. Sarah says that she is comfortable spending up to \$25 for the meal, while Laura says she is comfortable spending up to \$20 for the meal. They agree to split the bill evenly. If they spend a total that both are comfortable with, what is the range of money they spend on the meal in total (that is, what is the range of the possible bills?)? (Assume they pay something for the meal!)

Problem 2.26 Elsa is planning to make a rectangular garden surrounded by a fence to keep the rabbits out. She wants the garden to be 4 feet longer than it is wide. The fence she will be using costs $1.25 per foot. If she is willing to spend up to $100 on the fence, what are the possible values for the width of the fence in feet?

Problem 2.27 Ricky is on a long road trip. He wants to drive at least 400 miles in the next 8 hours. Some of the time he will be able to drive 60 miles per hour, and some of the time he will have to drive slower at 30 miles per hour. What range of values for the number of hours Ricky drives at 60 miles per hour will allow him to accomplish his goal?

Problem 2.28 Charlie and his two very competitive friends Danny and Frank played a video game at the arcade. Charlie played the game first. Danny played next and beat Charlie's score by one point. Frank played last and beat Danny's score by one point. If the total of all the scores is at least 500, what are the possible scores Charlie got for the game? You can assume all the scores are whole numbers.

Problem 2.29 Solve the inequality $p^2 - 4 < 21$.

Problem 2.30 Find all s such that $s^2 - 10 > 71$ and $2s + 7 < s + 3$.

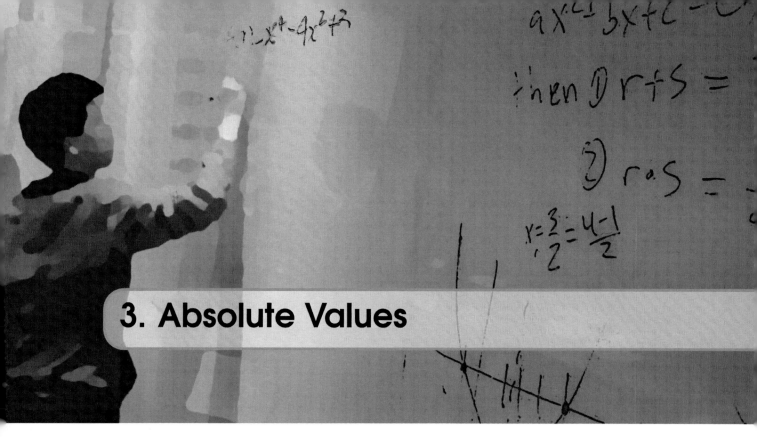

3. Absolute Values

Basic Definitions and Properties

- The absolute value of a number x, denoted $|x|$, is the distance from 0 to x on the number line.
- Properties of Absolute Values
  - $|x| \geq 0$ for all x, with $|x| = 0$ if and only if $x = 0$.
  - $|x| = |-x|$ for all x.
  - $|x \times y| = |x| \times |y|$.
  - Caution: It is not true that $|x + y| = |x| + |y|$!
- If x is positive, then $|x| = x$. If x is negative, then $|x| = -x$. We can write this as a formula as:
$$|x| = \begin{cases} x, & \text{if } x \geq 0; \\ -x, & \text{if } x < 0. \end{cases}$$
- The standard method to deal with absolute values is case analysis: solve in intervals where the expressions inside the absolute value does not change signs.

3.1 Example Questions

Problem 3.1 Find the value of the following absolute value expressions.

(a) $|-3|+|4|$

(b) $-|3|-|-5|$

(c) $|2||-4+8|-|-4||5-3|$

Problem 3.2 Plug in the variables in the expressions.

(a) $|2x-4|+|-x|, x=6$

(b) $3|-2x+1|-|x+8|, x=-5$

Problem 3.3 Solve the following absolute value equations.

(a) $|x|=10$

(b) $|-3+x|=8$

(c) $|-8x+4|=-3$

Problem 3.4 Solve the following absolute value inequalities and graph the solutions on the number line.

(a) $|x| \leq 3$

(b) $|x| > 2$

(c) $|x+2| - 1 \leq 3$

Problem 3.5 Solve the following absolute value inequalities.

(a) $3|x-2| + 3 < 12$

(b) $|x+4| + 4 < 12$

(c) $2|3 - 2x| - 6 \geq 18$

Problem 3.6 Solve the following absolute value equations.

(a) $|x| + 3 = |2x|$

(b) $|x+3| = |x|$

Problem 3.7 Solve the following equations with nested absolute values.

(a) $|3+|x+2||=4$

(b) $|x-|2x+1||=3$

Problem 3.8 Find all values of x, y and z such that $|x-6|+|y^2-4|+|z+8|=0$

Problem 3.9 Solve the following absolute value equations.

(a) $|x+1|+|3(x+1)|=8$

(b) $4|x-6|-|7x-42|\geq-3$

Problem 3.10 Write an absolute value inequality that is equivalent to the compound inequality.

(a) $-3\leq x\leq 3$

(b) $-2\leq x\leq 4$

(c) $\frac{3}{2}<x<\frac{9}{2}$

3.2 Quick Response Questions

Problem 3.11 What is the value of $|4| + |-3| - |8|$

Problem 3.12 What is the value of $|-5||7-6| + |4|$?

Problem 3.13 What is $|x+4| - |y-2|$ equal to when $x = -2$ and $y = 4$?

Problem 3.14 What expression is the same as $4|x+3|$?

(A) $4x + 12$
(B) $|4x + 3|$
(C) $|4x + 12|$
(D) $4x + 3$

Problem 3.15 What expression is not the same as $|6x - 8y|$?

(A) $|8y - 6x|$
(B) $2|3x - 4y|$
(C) $6|x| - 8|y|$
(D) $|-2||3x - 4y|$

Problem 3.16 What is the sum of the solutions to the equation $|x| = 9$?

Problem 3.17 How many solutions are there to the equation $|-x| = -2$?

Problem 3.18 How many solutions are there to the equation $|3 - 2x| = 0$?

Problem 3.19 How many integer solutions are there to the inequality $|3x| \leq 9$?

Problem 3.20 What is the smallest positive integer that is a solution to the inequality $|3x - 4| > 12$

3.3 Practice Questions

Problem 3.21 Calculate the following if $a = -2$ and $b = 5$.

(a) $|2a + b| - |a|$.

(b) $\left| \dfrac{|a+b-5|}{5-b-a} \right|$.

Problem 3.22 Solve the following absolute value equations.

(a) $|x - 8| = 6$

(b) $|x + 4| = 3$

Problem 3.23 Solve the following absolute value equations.

(a) $|9 + 7x| = 4$

(b) $|3x + 8| = -8$

Problem 3.24 Solve the following absolute value inequalities and graph their solution in the number line.

(a) $|2x| \leq 8$

(b) $|x-3| > 2$

Problem 3.25 Solve the following absolute value inequalities.

(a) $8 - |x+5| > 14$

(b) $4|x+9| - 7 < 12$

Problem 3.26 Solve for x: $|3x| + 9 = |6x|$.

Problem 3.27 Solve for x: $|x-2| = |x+2|$

Problem 3.28 Solve for x: $||x+4| - 4| = 4$

Problem 3.29 Solve for x: $|x-2| = |4x-8| - 3| - 3x+6|$

Problem 3.30 Solve for x $3|9-6x| \geq |2x-3| + |6x-9|$

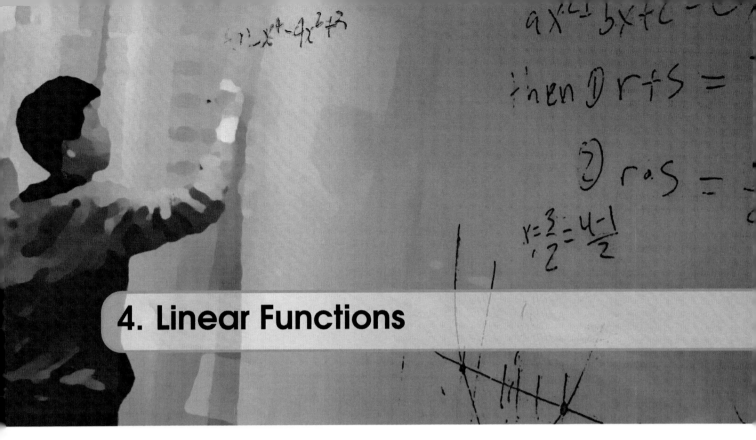

4. Linear Functions

Relations and Functions

- **Relations**
    - A relation is a relationship of inputs with outputs. Every input has an output, but it is possible for the same input to have multiple outputs.
    - It is often useful to use tables or graphs to help visualize relations.
- **Functions**
    - A function is a specific type of relation where each input has exactly one output.
    - In addition to tables or graphs, functions can often be described by equations.

Coordinate Plane

- It is often helpful to label points in a grid. Consider a horizontal number line, called the x-axis, combined with a vertical number line, called the y-axis as in the diagram below.

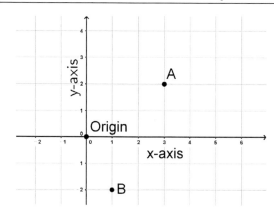

- For any point on the plane, its horizontal distance from the origin is called the *x*-coordinate, its vertical distance from the origin is called the *y*-coordinate, and the point is labeled with coordinates (x, y).
- For example, the point *A* in the diagram above has coordinates $(3, 2)$ and the point *B* is $(1, -2)$. The origin has coordinates $(0, 0)$.
- **Quadrants**
  - Points with $x > 0$ and $y > 0$ are in the first quadrant, or Quadrant I.
  - Points with $x < 0$ and $y > 0$ are in the second quadrant, or Quadrant II.
  - Points with $x < 0$ and $y < 0$ are in the third quadrant, or Quadrant III.
  - Points with $x < 0$ and $y < 0$ are in the fourth quadrant, or Quadrant IV.
  - Note: Points on the *x*-axis or *y*-axis are NOT in one of the quadrants.
- Linear equations and linear inequalities can often be graphed in the coordinate plane to help understand solution sets.

Slopes

- Slope between two points *A* and *B* is given by the equation $\dfrac{y_A - y_B}{x_A - x_B}$.
- **Special Cases**
  - Note if both points have the same *y*-coordinate, the slope is 0. In this case, the points are horizontal to each other.
  - Note if both points have the same *x*-coordinate, the slope is undefined because we cannot divide by 0. In this case, the points are vertical to each other. In certain cases, it makes sense to think of this slope as being 'infinite'.

Lines

- The graphs of linear equations or functions are called lines.
- Lines are often written in one of three forms:
  - Standard form: $Ax + By = C$.

      ○ Slope-intercept form: $y = mx + b$ (m is the slope, b is the y-intercept).
      ○ Point-slope form: $y - y_A = m(x - x_A)$ for a line with slope m going through point A.

4.1 Example Questions

Problem 4.1 On a single coordinate plane, plot the following points: $A = (0, 1)$, $B = (2, -3)$, $C = (5, 1)$, $D = (-1, 5)$, $E = (-3, -3)$. For each of the points, state explain its position in the plane (which axis it is on, which quadrant it is in, etc.).

Problem 4.2 Graph the following equations on a coordinate plane.

(a) $y = 2x - 1$.

(b) $y = x^2$.

(c) $y = |x|$

Problem 4.3 Find the slope between the two given points.

(a) $(2, 3)$ and $(4, 9)$.

(b) $(-2, 4)$ and $(6, 4)$.

(c) $(-3, 5)$ and $(1, 0)$.

(d) $(2, 0)$ and $(2, -1)$.

Problem 4.4 For each of the following relations, classify the following as (i) a linear function, (ii) a function that is not linear, or (iii) not a function.

(a)

x	1	2	3	5	8
$f(x)$	2	3	4	6	9

(b)

x	-2	-1	0	1	4
$f(x)$	-4	-1	0	-1	-16

(c)

x	1	1	2	2	4	4
$f(x)$	-1	1	-2	2	-4	4

Problem 4.5 Write a linear equation in slope-intercept form for each of the following.

(a) A linear equation with slope -2 containing the point $(0, 2)$.

(b) A linear equation with slope 4 containing the point $(-2, 7)$.

Problem 4.6 Write a linear equation in point-slope form for each of the following.

(a) A linear equation with slope -2 containing the point $(3,7)$.

(b) A linear equation containing the points $(2,3)$ and $(8,2)$.

Problem 4.7 Write a linear equation in standard form for each of the following.

(a) The linear equation $y = 2x + 4$.

(b) The linear equation with x-intercept 2 and y-intercept 4.

Problem 4.8 Graph the following linear inequalities.

(a) $y > -\dfrac{1}{2}x + 2$.

(b) $2x - 3y < 12$

Problem 4.9 Graph the following equations with absolute values.

(a) $y = |x|$.

(b) $y = 2|x - 2|$.

(c) $|2x - y| = 1$.

Problem 4.10 Graph the following inequalities with absolute values.

(a) $y > |x + 1|$.

(b) $|x + y| < 2$.

4.2 Quick Response Questions

Problem 4.11 In which quadrant is the point $(-2,3)$?

(A) Quadrant I
(B) Quadrant II
(C) Quadrant III
(D) Quadrant IV

Problem 4.12 Find the slope between the points $(4,2)$ and $(1,8)$. Round your answer to the nearest tenth if necessary.

Problem 4.13 What is the slope of the line $y = 3x - 4$?

Problem 4.14 What is the slope of the line $x = 4$?

(A) 0
(B) 1
(C) Undefined because there is no change in x.
(D) Undefined because there is no change in y.

Problem 4.15 What is the slope of the line $5x + 2y = 10$? Round your answer to the nearest tenth if necessary.

Problem 4.16 Which of the following relations does NOT describe a function?

(A) $f(x) = 2x + 4$
(B) Input: Date (including month, day, and year), Output: Day of week on that day
(C) $y = x^2$
(D) Input: Date (including month, day, and year), Output: People born on that day

Problem 4.17 What is the equation of the line $y = -3x + 2$ in standard form?

(A) $y - 2 = -3x$
(B) $3x + y = 2$
(C) $3x - y = 2$
(D) $3x + y = -2$

Problem 4.18 Written in point-slope form, what is the equation of the line with slope -2 what contains the point $(5, 2)$?

(A) $y - 2 = -2(x - 5)$
(B) $y = -2x + 2$
(C) $y = -2x + 12$
(D) $5x + 2y = -2$

Problem 4.19 Does the point $(2, 3)$ satisfy the inequality $|x - y| \le 2$?

Problem 4.20 Consider the graph of a linear inequality below:

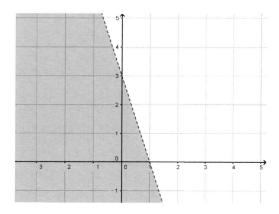

Which of the following describes the equation for the shaded region?

(A) $y > -3x + 3$
(B) $y \geq -3x + 3$
(C) $y < -3x + 3$
(D) $y \leq -3x + 3$

4.3 Practice Questions

Problem 4.21 On a single coordinate plane, plot the following points: $A = (3,1)$, $B = (-3,0)$, $C = (-4,-2)$, $D = (2,0)$. For each of the points, state explain its position in the plane (which axis it is on, which quadrant it is in, etc.).

Problem 4.22 Graph the equation $y = |2x - 4|$ by plotting points.

Problem 4.23 Find the slope between the given points.

(a) $(3,2)$ and $(-8,-2)$.

(b) $(10,2)$ and $(10,-2)$.

Problem 4.24 Suppose the points $(2,3)$, $(5,7)$ and $(-1,c)$ all satisfy the same linear equation. What is c?

Problem 4.25 Consider the line through the points $(-1,0)$ and $(1,4)$. Write the equation of this line in slope-intercept form. Try to do this without finding any other forms first. Explain your method.

Problem 4.26 Consider the line through the points $(-2,2)$ and $(2,-2)$. Write the equation of this line in point-slope form. Try to do this without finding any other forms first. Explain your method.

Problem 4.27 Consider the line through the points $(-1,0)$ and $(0,3)$. Write the equation of this line in standard form. Try to do this without finding any other forms first. Explain your method.

Problem 4.28 Graph the inequality $y + 2 \leq \frac{-(x+2)}{2}$.

Problem 4.29 Graph the equation $y = -|x+1|$.

Problem 4.30 Graph the inequality $y \geq |x| - 2$.

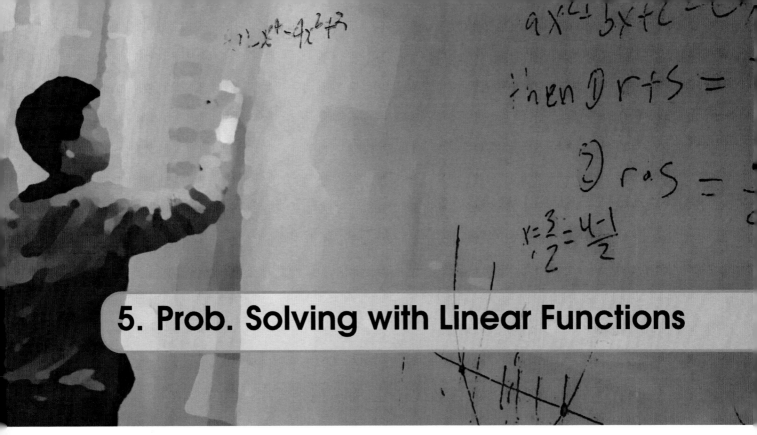

5. Prob. Solving with Linear Functions

Slopes Review

- Slope between two points A and B is given by the equation $\dfrac{y_A - y_B}{x_A - x_B}$.
- Special Cases
  - If both points have the same y-coordinate, the slope is 0. In this case, the points are horizontal to each other.
  - If both points have the same x-coordinate, the slope is undefined because we cannot divide by 0. In this case, the points are vertical to each other. In certain cases, it makes sense to think of this slope as being 'infinite'.

Lines Review

- The graphs of linear equations or functions are called lines.
- Lines are often written in one of three forms:
  - Standard form: $Ax + By = C$.
  - Slope-intercept form: $y = mx + b$ (m is the slope, b is the y-intercept).
  - Point-slope form: $y - y_A = m(x - x_A)$ for a line with slope m going through point A.

Parallel and Perpendicular Lines

- Two lines with the same slope are called *parallel*.
- Two lines whose slope multiply to -1 are called *perpendicular*. Note this is equivalent to saying the slopes are opposite reciprocals.

$\circ$ For example, $y = 2x + 3$ and $y = -\dfrac{1}{2}x - 2$ are perpendicular lines.

Variation

- **Direct Variation**
  - $\circ$ "y varies directly as x" or "y is directly proportional to x" if there is a number k (called the constant of variation) so that $y = k \times x$.
  - $\circ$ For example, if George drives a car at 60 mph, then the distance D George drives is directly proportional to T, the time that he drives. Here the constant of variation is 60.
- **Inverse Variation**
  - $\circ$ "y varies inversely as x" or "y is inversely proportional to x" if there is a number k (called the constant of variation) so that $x \times y = k$. Note this is equivalent to $y = \dfrac{k}{x}$.
  - $\circ$ For example, if Sarah has 100 math problems to complete, then the time T it takes her to complete the problems is inversely proportional to N, the number she can do per hour. Here the constant of variation is 100.
- **Joint Variation**
  - $\circ$ "z varies jointly as x and y" or "z is jointly proportional to x and y" if there is a number k (called the constant of variation) so that $z = k \times x \times y$.
  - $\circ$ For example, the area A of a rectangle is jointly proportional to L and W, the length and width of the rectangle. Here the constant of variation is 1.

5.1 Example Questions

Problem 5.1 Find the equations for the following lines.

(a) The line with x-intercept 4 and slope -2.

(b) The line containing the points $(-2, -4)$ and $(2, 6)$.

Problem 5.2 For each of the following pairs of lines determine whether they are parallel, perpendicular, or neither.

(a) $y = 2x + 3$ and $y = -\frac{1}{2}x + 2$

(b) $y = -\frac{3}{4}x + 4$ and $3x + 4y = 24$.

(c) The line containing the points $(2,0)$ and $(-1,3)$ and the line $y = x - 4$.

Problem 5.3 Find equations for the lines described below.

(a) The line containing the point $(0,2)$ that is parallel to the line containing the points $(1,2)$ and $(3,4)$.

(b) The line containing the origin that is perpendicular to the line $2x + y = 4$.

Problem 5.4 Write equations describing the following situations. If they describe a direct varation, inverse varation, or joint variation, say which variation is described.

(a) Larry has \$400 to spend on collectible action figures. If the price of each action figure is P, write an equation describing N, the number of action figures Larry can buy.

(b) William gets paid \$8 an hour for his work. If he works for T hours, write an equation describing P, the total payment William gets for his work.

(c) Cary drives a semi and gets paid $0.75 per mile he drives. If he drives for T hours at R miles per hour, write an equation describing P, the total payment Cary gets for driving.

(d) Patricia sells printers at the local office supplies store. He gets paid a base salary of $200 per week, and earns 4% commission on her sales. If she makes a total of S dollars in sales this week, write an equation describing W, Patricia's total wage for the week.

Problem 5.5 Fill in the missing values in the tables below.

(a)

x	y
0.5	
1	300
	10
600	

where y is inversely proportional to x.

x	y
$\frac{1}{3}$	$\frac{1}{4}$
(b)	6
12	
20	

where y is directly proportional to x.

Problem 5.6 A football is thrown downward from the top of a building. Its velocity is 38 feet per second after 1 second and 70 feet per second after 2 seconds.

(a) The velocity v of the football can be given using a linear equation in terms of the time t. What is this equation?

(b) What was the initial velocity of the ball?

(c) What is the velocity of the football after 4.5 seconds?

(d) After how many seconds is the ball traveling 100 feet per second? Round your answer to the nearest tenth if necessary.

Problem 5.7 William wants to enclose a rectangular portion of his backyard using a fence. For the East and West sides of his enclosure he will use one type of fence that costs $8 per yard. For the other two sides (North and South) he uses a fence that costs $5 per yard. William wants to spend at most $200 on his fence.

(a) Write an inequality in terms of the width W (the length of the East and West sides) and the length L (the length of the North and South sides) to help William make sure the size he chooses fits his budget.

(b) William ends up deciding that he wants the length L to be twice the width W. If the dimensions of the rectangular enclosure are integers, what is the area of the largest enclosure William can build?

Problem 5.8 Consider the data given below.

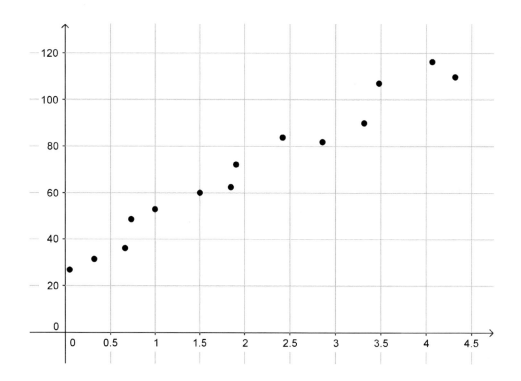

Model the data using a line that has a slope and y-intercept that are both multiples of 10.

Problem 5.9 Piper has a pepper farm. Each summer, before growing season, she uses some horses to help prepare the field. Last year she used 4 horses and it took her 75 minutes. This year she used 6 horses and it took her 50 minutes. Next year she plans to use 8 horses.

(a) Suppose the time it takes is inversely proportional to the number of horses used. How many minutes will it take Piper to prepare the pepper field next year?

(b) Suppose the time it takes follows a linear equation based on the number of horses used. How many minutes will it take Piper to prepare the pepper field next year?

(c) Which of the models (described in parts (a) and (b)) do you think is more accurate? Explain your answer.

Problem 5.10 Lauren studies the growth of plants at a laboratory. Lauren hypothesizes that the height of the plants she studies is directly proportional to the number of days of growth with a growth rate of 1.5 inches per day.

(a) Write an equation describing Lauren's hypothesis. Use H to represent the plant's height in inches and D to represent the number of days.

(b) Before looking at her data, Lauren decides that she will accept her hypothesis as long as the height of the plants is within 1 inch (inclusive) of her model. Write an inequality using absolute values that the data must satisfy for Lauren to accept her hypothesis.

5.2 Quick Response Questions

Problem 5.11 Which one of the lines below is NOT parallel to the rest?

(A) $y = 3x - 2$
(B) $6x - 2y = 1$
(C) $3x + y = 4$
(D) $y - 2 = 3(x + 2)$

Problem 5.12 Which one of the lines below is perpendicular to the line $y = 2x + 4$?

(A) $y = 2x - 4$
(B) $y = -2x + 4$
(C) $y = \dfrac{1}{2}x + 4$
(D) $y = -\dfrac{1}{2}x - 4$

Problem 5.13 Is it possible for a line to be parallel to the line $y = 2x + 3$ and perpendicular to the line $y = \dfrac{1}{2}x - 2$?

(A) No, because it is impossible to find a y-intercept for such a line.
(B) No, because it is impossible to find a slope for such a line.
(C) Yes, multiple examples of lines exist.
(D) Yes, but only one example exists.

Problem 5.14 Is it possible for a line to share the y-intercept of the line $y = 2x + 3$ and be perpendicular to the line $y = \dfrac{1}{2}x - 2$?

- (A) No, because it is impossible to find a y-intercept for such a line.
- (B) No, because it is impossible to find a slope for such a line.
- (C) Yes, multiple examples of lines exist.
- (D) Yes, but only one example exists.

Problem 5.15 Which of the following equations describes a situation where y varies inversely as x?

- (A) $xy = 4$
- (B) $y = \dfrac{x}{3}$
- (C) $x = \dfrac{y}{2}$
- (D) Multiples choices above work.

Problem 5.16 Which of the following equations describes a situation where y varies directly as x?

- (A) $xy = 4$
- (B) $y = \dfrac{x}{3}$
- (C) $x = \dfrac{y}{2}$
- (D) Multiples choices above work.

Problem 5.17 A ball is dropped from a roof. After 1 second its velocity is 32 ft/sec. If the velocity is directly proportional to the time, what is the velocity after 3 seconds? Give your answer in ft/sec.

Problem 5.18 Suppose that z varies jointly with x and y. Further, $z = 60$ when $x = 4$ and $y = 6$. What is z when $x = 8$ and $y = 2$?

Problem 5.19 Gary has a huge oak tree in his front yard. During Fall he wanted to model the rate at which leaves fell off the tree using a linear model. On the first day of Fall, 3 leaves fell of the tree. On the sixth day of Fall, 13 leaves fell. If Gary's model line contains both of these data points, how many leaves does his model estimate will fall on the tenth day of Fall?

Problem 5.20 A company sells fences that are 6 feet tall. A customer wants to build a fence in a rectangular field that is L feet long and W feet wide. Which inequality below describes a total surface area of the fence being sold to the customer that is greater than 600 square feet.

(A) $6LW > 600$

(B) $12L + 12W > 600$

(C) $6L + 6W > 600$

(D) $12LW > 600$

5.3 Practice Questions

Problem 5.21 Find the equations for the following lines. Give your answers in standard form.

(a) The line with slope -4 containing the point $(1, -4)$.

(b) The line containing the points $(-3, 0)$ and $(0, 4)$.

Problem 5.22 Suppose $y = mx + b$ and $y = bx + m$ are perpendicular lines. Find these lines if both have an x-intercept of $x = 1$.

Problem 5.23 Find the equation of the line that shares the x-intercept of $y = 3x - 5$ but is perpendicular.

Problem 5.24 Recall 5.4. Write your own real world examples, one each for direct, inverse, and joint variations.

Problem 5.25 David is told that x and y are proportional and knows that $x = 60$ when $y = 40$. However, David is unsure whether they are directly or inversely proportional. If $x = 24$, what are all possible values of y?

Problem 5.26 Suppose a baseball pitch leaves the pitchers hand traveling at 90.90 mph. Using a high speed camera, the speed of the baseball was 90.20 mph 0.2 seconds later. If the baseball continues to slow down at a linear rate due to air resistance, what will be the speed of the pitch when it hits the catcher's glove 0.5 seconds after it was thrown? Round your answer to the nearest hundredth if necessary.

Problem 5.27 Bill and Rick are two friends. Bill has a job that earns $10 an hour and Rick has a job that earns $12 an hour. Their goal this week is to earn $800. Assume Bill works x hours and Rick works y hours. Write a linear inequality for when the two friends accomplish their goal and solve your inequality for y.

Problem 5.28 Consider the collection of data found below.

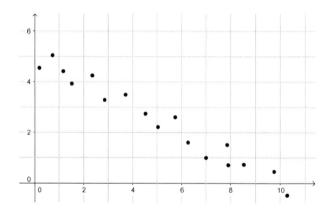

If we assume that the data can be modeled by a linear equation with integer intercepts, find an equation for the line.

Problem 5.29 Recall 5.9 Piper's brother Tommy has his own farm where he grows tomatoes. He also uses horses on his farm to help plant tomato plants. The total number of plants Tommy can plant is jointly proportional to the number of horses he has and the number of hours he works. Working for 2 hours with 4 horses Tommy can plant 300 tomato plants. How many tomato plants can Tommy plant if he works for 6 hours with 3 horses?

Problem 5.30 Recall 5.10. Lauren collected data from 5 plants over the course of 10 days. The height's of the plants (in inches) over the first 10 days of growth are summarized below.

	Plant 1	Plant 2	Plant 3	Plant 4	Plant 5
Day 1	1.1	2	1.4	1.4	1.9
Day 2	2.2	4	2.9	2.5	3.2
Day 3	3.3	6	4.6	4.5	4.6
Day 4	4.4	8	6.2	5.5	6.5
Day 5	5.5	9	7.9	6.8	8.5
Day 6	6.6	10	9.4	8.6	10.4
Day 7	7.7	11	10.8	9.9	11.6
Day 8	8.8	12	12.2	11.4	12.9
Day 9	9.9	13	13.2	13.2	14.7
Day 10	11	14	14	14.3	16.1

What percentage of the 50 data points above fit within 1 inch (inclusive) of Lauren's model that $H = 1.5D$?

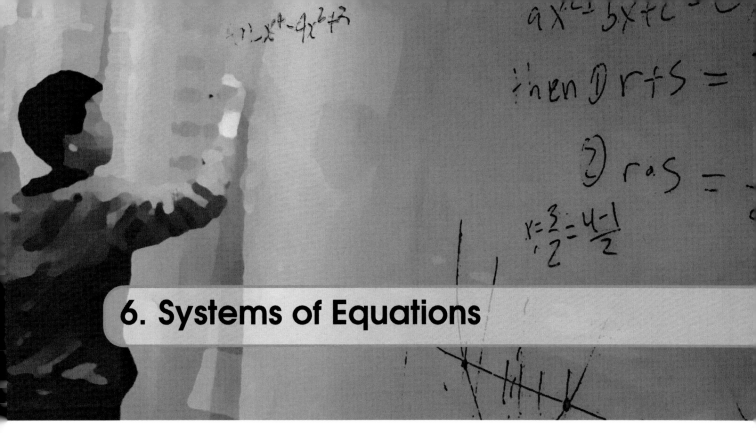

6. Systems of Equations

Systems of Equations or Inequalities

- Thus far we have discussed solving equations and inequalities with linear equations.
- What happens if we want multiple equations to be true at once? This is called a system of equations.
- Many methods exist to solve systems of equations, the two most common that we will see examples of today are
  - ○ Substitution
  - ○ Elimination

6.1 Example Questions

Problem 6.1 Graph the following pairs of lines on one coordinate plane. Do they intersect? If so, where?

(a) (i) $y = 2x - 3$, (ii) $y = 4x + 3$

(b) (i) $2x + y = 3$, (ii) $4x + 2y = -4$

(c) (i) $x - 2y = 4$, (ii) $y + 1 = \dfrac{1}{2}(x - 2)$

Problem 6.2 Solve the following systems of equations using substitution.

(a) (i) $y = 2x - 3$, (ii) $2y - x = 3$

(b) (i) $x - y = 3$, (ii) $3x - 2y = 4$

Problem 6.3 Solve the following systems using elimination.

(a) (i) $x + y = 10$, (ii) $x - y = 8$

(b) (i) $6a + 5b = 7$, (ii) $3a + 5b = 1$

Problem 6.4 Solve the following equations using elimination.

(a) (i) $4x + y = 11$, (ii) $5x - 2y = 4$.

(b) (i) $\dfrac{1}{2}x - \dfrac{2}{3}y = \dfrac{7}{3}$, (ii) $\dfrac{3}{2}x + 2y = -25$

Problem 6.5 Graph the solution sets to each of the inequalities below.

(a) (i) $y \geq x - 3$, (ii) $y > -x + 1$.

(b) (i) $y \leq 2x + 3$, (ii) $y > -3x - 2$.

Problem 6.6 Graph the solution sets for the following.

(a) (i) $y \geq |x| - 3$, (ii) $x - y \geq -1$.

(b) (i) $y \leq 2x - 1$, (ii) $4x - 2y > -2$

Problem 6.7 Solve the system if (i) $x + y + z = 10$, (ii) $x = 4y$, and (iii) $z = -y$.

Problem 6.8 Solve the following systems of equations with three variables and three unknowns.

(a) (i) $x + y + z = 3$, (ii) $x + 2y + 3z = 4$, and (iii) $x + 2y + 4z = 5$.

(b) (i) $6x + 2y - z = 3$, (ii) $4x + 3y + z = 7$, and (iii) $z = x + 3$.

Problem 6.9 Consider the equations: (i) $x + y - z = 4$, (ii) $2x + 2y - 2z = 4$, and (iii) $x + y - 2z = 4$. Which pairs of the pairs, (i) & (ii), (i) & (iii), (ii) & (iii), have solutions? Does the system of all three equations have a solution?

Problem 6.10 Solve the following systems using some of the techniques from today.

(a) (i) $x^2 - y^2 = 4$, (ii) $x^2 + y^2 = 4$.

(b) (i) $|2x| + y = x + y + 2$, and (ii) $y = 6x - 4$.

6.2 Quick Response Questions

Problem 6.11 Is $(1,2)$ a solution to the system (i) $2x+y=4$, (ii) $3x-y=6$.

Problem 6.12 How many times do the graphs of $2x+3y=5$ and $2x-3y=7$ intersect?

(A) 0
(B) 1
(C) 2
(D) Infinitely many times

Problem 6.13 How many times do the graphs of $4x-y=2$ and $y=2(2x-1)$ intersect?

(A) 0
(B) 1
(C) 2
(D) Infinitely many times

Problem 6.14 How many times do the graphs of $y=x^2$ and $y=4$ intersect?

(A) 0
(B) 1
(C) 2
(D) Infinitely many times

Problem 6.15 What is the result of adding the equations $3x + y = 9$ and $4x - y = 15$?

(A) $7x = 24$
(B) $7x + 2y = 24$
(C) $x - 2y = -6$
(D) $7x = 6$

Problem 6.16 If $x + y = 10$ and $x - y = 4$, what is the value of x?

Problem 6.17 If $4x + 3y = 8$ and $4x + 8y = 13$, what is the value of y?

Problem 6.18 Suppose the system (i) $2x - y = 1$, (ii) $ax + y = 13$ has solution $(2, 3)$. What is a?

Problem 6.19 If $x + 2y + 3z = 2$ and $x + 2y + z = 8$, what is the value of z?

Problem 6.20 In the system (i) $x - y = 3$, (ii) $y = 4z$, (iii) $x + y + z = 21$, what is z?

6.3   Practice Questions

Problem 6.21 Graph the equations $y = x^2$ and $y = x + 2$ by plotting points. Where do they intersect?

Problem 6.22 Solve the system of equations (i) $x = 8y$, (ii) $2x + 3y = 38$ by substitution.

Problem 6.23 Solve the system of equations (i) $\frac{2}{3}x - \frac{1}{2}y = 12$, (ii) $\frac{5}{6}x - \frac{1}{2}y = 14$ by elimination.

Problem 6.24 Solve the system (i) $x - y = 100$, (ii) $0.2x + 0.06y = 150$.

Problem 6.25 Graph the solution set to $x + y \leq 3$ and $x > 2$.

Problem 6.26 Graph the solution set to $x - 3y \leq 6$ and $y < -|x - 2|$.

Problem 6.27 Solve the system of equations (i) $x + y + z = 9$, (ii) $y + z = 7$, and (iii) $z = 3$.

Problem 6.28 Solve the system of equations (i) $x + y + z = 4$, (ii) $2x + y + z = 6$, and (iii) $-x + y - z = 6$.

Problem 6.29 Solve the system of equations (i) $x - y = 3$, (ii) $y + z = 8$, and (iii) $2x + 2z = 7$.

Problem 6.30 Solve the system of equations (i) $\frac{1}{2}\sqrt{x} - 2y = 4$, (ii) $\frac{1}{2}\sqrt{x} + 2y = 6$.

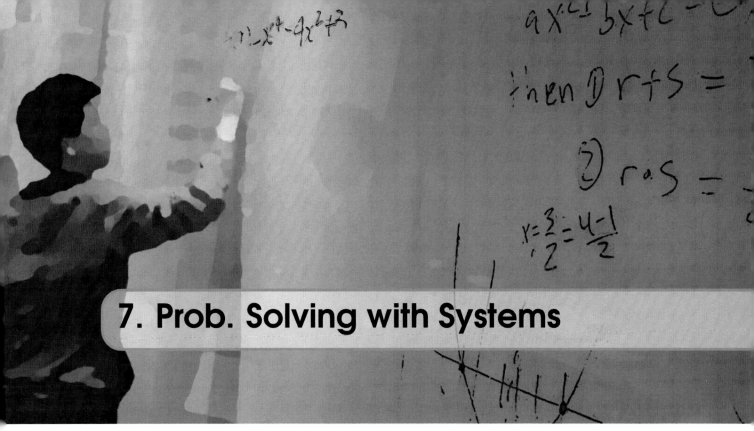

7. Prob. Solving with Systems

Review of Systems of Equations

- Solving multiple equations at once is called solving a system of equations.
- Recall the two main methods for solving systems:
  - Substitution
  - Elimination

7.1 Example Questions

Problem 7.1 Systems of Equations Review

(a) Solve (i) $x - y + z = 5$, (ii) $y - z = 3$, (iii) $x = 4$.

(b) Solve (i) $3x - 4y + z = 8$, (ii) $2x + 4y + z = -1$, (iii) $x - 2y + 3z = 6$.

(c) Solve (i) $2|x| + y = 5$, (ii) $|x| - y = 1$.

Problem 7.2 Solve the following questions (i) using a system of equations, (ii) without using a system of equations. Compare the two methods.

(a) Suppose there are chicken, rabbits, and sheep on a farm. There are 70 heads in total and 220 feet. If there are the same number of rabbits and sheep, how many chickens, rabbits, and sheep are on the farm?

(b) Some chickens and rabbits have a total of 100 feet. If each chicken was exchanged for a rabbit, and each rabbit was exchanged for a chicken, there would be a total of 86 feet. How many chickens are there? How many rabbits?

Problem 7.3 Solve the following questions (i) using a system of equations, (ii) without using a system of equations. Compare the two methods.

(a) Sami and Rajan practice running together. If Sami starts to run after Rajan runs for 10 meters, then it will take Sami 5 seconds to catch up with Rajan. If Sami starts to run after Rajan runs for 2 seconds, then it will take Sami 4 seconds to catch up with Rajan. How fast can each person run?

(b) When two teams A and B work together, it takes 18 days to get a job completed. After team A works for 3 days, and team B works for 4 days, only $\frac{1}{5}$ of the job is done. How long does it take for team A alone to complete the job? For team B alone?

Problem 7.4 George starts to build his investment portfolio. He has $10,000 divided between stocks, bonds, and savings. He earns a 10% return on stocks, 6% on bonds, and 1% on savings. His total return was $520. If his total investment in stocks equaled his total investment in bonds, how much did George keep in savings?

Problem 7.5 Wilson has a total of 60 chickens and rabbits on a farm. One day he buys some number of chickens so that he has the same number of chickens as rabbits. The next day he buys the same number of chickens again. At this point he has 40 more rabbit legs than chicken legs. How many rabbits does Wilson have?

Problem 7.6 There are 50 bugs in total, made up of spiders (8 legs, no wings), houseflies (6 legs, 1 pair of wings), and dragonflies (6 legs, 2 pair of wings). There is one housefly for every dragonfly leg. If every spider is exchanged with a dragonfly, there is one housefly for every pair of dragonfly wings. How many dragonflies are there?

Problem 7.7 Alice, Bob, and Eve are washing cars. In total they wash 300 cars. Alice washes a third as many cars as Bob and Eve combined. Eve washes half as many cars as Alice and Bob combined. How many cars does Bob wash?

Problem 7.8 Troy and Max work together to paint a fence. If Troy works alone for 5 days and Max works for 3 days, they can finish the entire fence. If Max works for 1 day and then they work together for 1 day, they can finish 1/4 of the fence. How long does it take Max to paint the fence alone?

Problem 7.9 Bill, Claire, and Drew went on a road trip over the weekend. In total they drove 500 miles over 10 hours. Bill drove at an average rate of 40 mph, Claire an average of 55 mph, and Drew an average of 65 mph. If Bill and Drew drove the same distance, how long many hours did each drive?

Problem 7.10 Farmer Billy has a farm where he raises chickens and rabbits. One Sunday Billy counted 500 legs on the farm. On Monday he traded all his rabbits for chickens (1 rabbit for 1 chicken). On Tuesday, he traded all his original chickens (not the new ones) so that he received 3 rabbits for every 2 chickens. On Wednesday Billy remarked that there were still 500 legs on the farm. How many chickens and how many rabbits did Billy start the week with?

7.2 Quick Response Questions

Problem 7.11 If $x = 4y$ and $2x + 3y = 33$, what is x?

Problem 7.12 If $3s + 3t = 9$ and $3s - t = 5$ what is s?

Problem 7.13 If 8 people can finish a task in 14 days, how many days would it take to do the same job with 16 people? Round your answer to the nearest day if necessary.

Problem 7.14 Melisa drove for 3 hours at a rate of 50 miles per hour and for 2 hours at 60 miles per hour. What was her average speed for the whole journey in miles per hour?

Problem 7.15 Clarence invested \$3500 in a stock that earned him a 5% return last year. If invested \$1500 in a different stock that earned him a 10% return last year. Overall, Clarence's \$5000 investement earned him a $K\%$ return, where K is a decimal rounded to the nearest tenth if necessary. What is K?

Problem 7.16 As of the end of the 2016 baseball season, the Cardinals have won 50% more World Series than the Oakland Athletics. Together the two teams have 15 wins. Let c and a denote the number of World Series wins for the Cardinals and Athletics respectively. If you used a system of equations to help solve this problem, which equation below would appear in the system?

(A) $a + 50 = c$
(B) $a = 1.5c$
(C) $c = 1.5a$
(D) $c = 15a$

Problem 7.17 William works at a local shoe store. He earns a bonus of $10 for every pair of shoes shoe he sells, and a bonus of $1 for each pair of socks. Which of the following represents William selling x shoes and y socks for a toal bonus of 255 dollars?

(A) $10x + y = 255$
(B) $x + 10y = 255$
(C) $10x + 255 = y$
(D) $10x + y = 225$

Problem 7.18 There are 60 animals on a farm, made up of x chickens and y rabbits. In total there are 180 legs. Which of the following equations is a true statement involving x and y?

(A) $4x + 2y = 180$
(B) $2x + 4y = 60$
(C) $x + y = 90$
(D) $x + 2y = 90$

Problem 7.19 $\angle A$ and $\angle B$ are supplementary angles. If $\angle B$ is $36°$ larger than A, what is angle B (measured in degrees)?

Problem 7.20 Paul is twice as old as his son Chris. Adding their ages gives 60. How old is Paul?

7.3 Practice Questions

Problem 7.21 Solve the system of equations (i) $2x - 4y + 5z = 8$, (ii) $2x - 4y + 3z = 2$, (iii) $x + 4y + z = 7$.

Problem 7.22 Lily spent $490 to buy 80 color pencils for her art class, including red, green. and blue colors. The red pencils cost $2 each, the green ones cost $5 each, and the blue ones cost $10 each. Suppose she bought the same number of green and blue pencils. How many of each type of pencils did she buy?

Problem 7.23 Ben and Jack can finish a task in 6 days working together. If Ben works alone for 5 days and then Jack takes over and works for 3 days, they finish $\dfrac{7}{10}$ of the work. How long would it take for each of them complete the task working alone?

Problem 7.24 Lauren invests $1000 in three different stocks, Stock A, Stock B, and Stock C. In one year, Stock A gives a 30% rate of return, Stock B a 5% rate of return, and Stock C a -10% rate of return (so Lauren lost 10% of the money she invested in Stock C). If Lauren earned $80 in total and the total gain (in dollars) of Stock B equaled the combined gain (and loss) from Stocks A and C, how much did Lauren invest in each stock?

Problem 7.25 There are some number of chickens and rabbits on a farm. There are a total of 200 legs on the farm. If there were 30 more chickens and 15 less rabbits, then there would be an equal number of legs. How many chickens and rabbits were there initially?

Problem 7.26 Starfish have 5 legs, beetles have 6 legs, and crabs have 10 legs. The local aquarium has 50 total animals of these three types. There are the same number of crab legs as there are starfish legs. If there are 320 total legs, how many animals are there of each type?

Problem 7.27 The four friends James, Paul, Wade, and Anthony harvest bananas. In total they harvest 2000 bananas. James harvests twice as many bananas as Paul, and also twice as much as Wade. If James and Anthony together harvest as much as Paul and Wade, how many bananas do each of the friends harvest?

Problem 7.28 Rob and Jon work together to paint a wall. If Rob works a total of 4 hours and Jon a total of 3 hours they can finish painting the wall. Alternatively, if Rob works 8 hours and Jon 1 hour they can also finish the wall. What is the ratio of the rate Rob works to the rate Jon works?

Problem 7.29 Recall 7.9 Bill, Claire, and Drew went for another 500 mile road trip. Bill was upset by how much more he had to drive on the previous road trip, so it was agreed that this time Bill and Drew would drive the same amount of time. Is this possible? Note: The road trip is still 10 hours long, and Bill, Claire, Drew still drive at speeds of $40, 55, 65$ mph respectively.

Problem 7.30 Peanuts cost $4 a pound, while cashews cost $6 a pound. George has a mixture of peanuts and cashews that cost $50. If George eats half of the peanuts in the mixure, the remaining peanuts will be worth $2 less than the cashews. How many pounds of cashews are in the mixture?

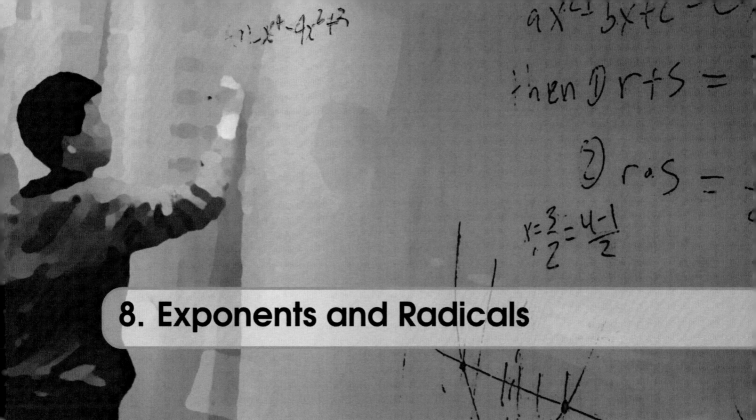

8. Exponents and Radicals

Exponents

- Exponents denote repeated multiplication. For example, $2^4 = 16$ because $2 \times 2 \times 2 \times 2 = 16$. Similarly $x^4 = x \times x \times x \times x$.
- The following are useful rules for exponents.
    - $x^0 = 1$ when $x \neq 0$.
    - $x^a \times x^b = x^{a+b}$.
    - $\dfrac{x^a}{x^b} = x^{a-b}$ or $\dfrac{x^a}{x^b} = \dfrac{1}{x^{b-a}}$.
    - $(x^a)^b = x^{ab}$.

Radicals

- The opposite of squaring a number is the square root. For example $\sqrt{4} = 2$ because $2 \times 2 = 4$. Note the symbol $\sqrt{}$, sometimes referred to as a radical, denotes taking the positive square root.
- The following are useful rules for exponents.
    - $\sqrt{x}$ is only defined for $x \geq 0$.
    - $\sqrt{a \times b} = \sqrt{a} \times \sqrt{b}$.
    - $\sqrt{\dfrac{a}{b}} = \dfrac{\sqrt{a}}{\sqrt{b}}$.
- Using these rules, we can often simplify radicals. A number is in simplest radical form if there are no squares left inside the radical. For example $\sqrt{8} = \sqrt{4 \times 2} = 2\sqrt{2}$ when written in simplest radical form. When simplifying fractions, radicals are not allowed in the denominator.

- Similarly we can define the cube, fourth, fifth, etc. roots. For example, the cube root of 8, written $\sqrt[3]{8} = 2$ because $2^3 = 8$.

8.1   Example Questions

Problem 8.1 Simplify the following expressions. You may assume any variables denote positive numbers.

(a) $(5x^2)^3$

(b) $\sqrt{169}$

(c) $\sqrt{16^3}$

(d) $\sqrt{9x^4}$

Problem 8.2 Write the following in simplest radical form.

(a) $\sqrt{32}$

(b) $3\sqrt{63}$

(c) $\sqrt{72}$

Problem 8.3 Write the following expressions in simplest radical form. You may assume that any of the variables used in the problems are positive.

(a) $\sqrt{20x^5}$

(b) $4\sqrt{243x^3y^5}$

Problem 8.4 Expand and simplify the following expressions.

(a) $3x^2(x+y+2)$.

(b) $2xy^2(x+3y)+2x^2(3x-y+5)$.

Problem 8.5 Adding and Subtracting Radicals

(a) $2\sqrt{8}-\sqrt{2}+3\sqrt{18}$

(b) $4\sqrt{24}+\sqrt{27}-\sqrt{54}$

Problem 8.6 Multiplying Radicals. Assume that any variables denote positive numbers.

(a) $\sqrt{2}\times\sqrt{8}$

(b) $2\sqrt{15}(r+\sqrt{5})$

(c) $\sqrt{3r}(\sqrt{r^3}+\sqrt{3})$

Problem 8.7 Exponents in Fractions

(a) $\dfrac{125x^3y^2}{5xy^2}$

(b) $\dfrac{80x^4y^2z^5}{120xy^4z^3}$

Problem 8.8 Simplifying Radicals in Fractions

(a) $\dfrac{\sqrt{18}-27}{3}$

(b) $\dfrac{\sqrt{6}-x\sqrt{3}}{\sqrt{3}}$

(c) $\dfrac{x^2-\sqrt{y}}{2\sqrt{7y}}$. Here you may assume $y > 0$.

Problem 8.9 Higher Power Roots

(a) Simplify $\sqrt[3]{108}$

(b) $\dfrac{\sqrt[3]{15}}{\sqrt[3]{64}}$

(c) $\dfrac{\sqrt[3]{10}}{\sqrt[3]{4}}$

Problem 8.10 Higher Power Radicals with Expressions. Assume any variables denote positive numbers.

(a) $2\sqrt[4]{243x^5 y^{12} z^3}$

(b) $\dfrac{4\sqrt{4m^2 n^4}}{\sqrt[4]{64m^4 n^2}}$

8.2 Quick Response Questions

Problem 8.11 Calculate $(\sqrt{25})^3$.

Problem 8.12 If we simplify $\sqrt{3} + 2\sqrt{3}$ we get:

(A) $3\sqrt{3}$
(B) $3\sqrt{6}$
(C) $2\sqrt{6}$
(D) $3\sqrt{9}$

Problem 8.13 Calculate $\sqrt{3} \times \sqrt{27}$ Round your answer to the nearest integer if necessary.

Problem 8.14 Which of the following is $\sqrt{150}$ in simplest radical form?

(A) $25\sqrt{5}$
(B) $15\sqrt{3}$
(C) $10\sqrt{15}$
(D) $5\sqrt{6}$

Problem 8.15 Which of the following is NOT equivalent to $\sqrt{144}$? ·

(A) 12
(B) $9\sqrt{16}$
(C) $6\sqrt{4}$
(D) $\sqrt{9} \times \sqrt{16}$

Problem 8.16 Calculate $\sqrt[3]{343}$.

Problem 8.17 A positive number that is a fifth power is called a perfect fifth power. For example, $1 = 1^5$ and $3^5 = 243$ are perfect fifth powers. How many perfect fifth powers are there less than or equal to $100,000$?

Problem 8.18 To write $\sqrt[3]{405}$ in simplest radical form we need to find the largest perfect cube C that is a factor of 405. What is C?

Problem 8.19 Simplify $\sqrt{\dfrac{9}{4}}$ and write your answer as a decimal. Round your answer to the nearest tenth if necessary.

Problem 8.20 Which of the following is equal to $\sqrt{11}(2\sqrt{22} + \sqrt{5})$?

(A) $22\sqrt{2} + 2\sqrt{55}$
(B) $2\sqrt{220} + \sqrt{55}$
(C) $22\sqrt{2} + \sqrt{55}$
(D) $2\sqrt{33} + 4$

8.3 Practice Questions

Problem 8.21 Simplify the following

(a) $(-3x^3)^2$

(b) $\sqrt{625}$

Problem 8.22 Write $\sqrt{250}$ in simplest radical form.

Problem 8.23 Write $\sqrt{48xy^3}$ in simplest radical form. You may assume that x and y are both positive.

Problem 8.24 Expand and simplify $3xy(x+2) - x(xy+2x)$.

Problem 8.25 Simplify the expression $\sqrt{18} + \sqrt{24} - \sqrt{8}$.

Problem 8.26 Distribute and simplify the terms $\sqrt{5x}(\sqrt{x} + \sqrt{5})$.

Problem 8.27 Simplify and reduce $\dfrac{72r^3s^5t^7}{2r^5st^3}$.

Problem 8.28 Simplify $\dfrac{2x - \sqrt{x}}{\sqrt{8x}}$.

Problem 8.29 Simplify $\dfrac{\sqrt[3]{30}}{\sqrt[3]{135}}$.

Problem 8.30 Simplify $\dfrac{\sqrt[3]{x^2 y}}{\sqrt[3]{xy^2}}$.

Review

- The distributive property states that $a(b+c) = ab + ac$ or $(a+b)c = ac + bc$.
- The following are useful rules for exponents.
  - $\sqrt{a \times b} = \sqrt{a} \times \sqrt{b}$.
  - $\sqrt{\dfrac{a}{b}} = \dfrac{\sqrt{a}}{\sqrt{b}}$.
  - $\sqrt{a}$ is in simplest radical form if there are no squares that are factors of a.

9.1 Example Questions

Problem 9.1 Review

(a) Write $\sqrt{756}$ in simplest radical form.

(b) Write $\sqrt[3]{756}$ in simplest radical form.

(c) Simplify $x^2(x+y+4)$

(d) Simplify $x\sqrt{20}(x+y\sqrt{2}+\sqrt{5})$

Problem 9.2 Expand the following.

(a) $(x+2)(x+4)$

(b) $(2x+y)(x+5)$

Problem 9.3 Expand the following expressions with radicals.

(a) $(2+\sqrt{2})(1-\sqrt{3})$

(b) $(\sqrt{2}-\sqrt{3})(\sqrt{10}-1)$

Problem 9.4 Expand the following. Note that these are very useful formulas!

(a) $(a+b)^2$

(b) $(a+b)(a-b)$

Problem 9.5 Expand the following, simplifying where possible.

(a) $(x^2 + 2)^2$

(b) $(\sqrt{2} + \sqrt{5})^2$

(c) $(4 + \sqrt{3})(4 - \sqrt{3})$

Problem 9.6 Simplify the following expressions.

(a) $\dfrac{11}{\sqrt{3} - 5}$

(b) $\dfrac{4\sqrt{21}}{\sqrt{3} + \sqrt{7}}$

Problem 9.7 Simplify $\dfrac{3 - \sqrt{y}}{3 + \sqrt{y}}$

Problem 9.8 Solve the following equations.

(a) $2x\sqrt{3} = \sqrt{6}$.

(b) $x + x\sqrt{2} = 4$.

Problem 9.9 Solve the following equations by squaring both sides.

(a) $3\sqrt{7} = \sqrt{-y}$.

(b) $4 + \sqrt{x+4} = 0$

(c) The square root of the sum of a number and 7 is 8. What is the number?

Problem 9.10 Expand the following.

(a) $(x+y+z)^2$

(b) $(\sqrt{2}+\sqrt{3}-1)^2$

9.2 Quick Response Questions

Problem 9.11 Which of the following is equal to $\sqrt{2}(\sqrt{2}-1)$?

(A) $2+\sqrt{2}$

(B) 1

(C) $4-\sqrt{2}$

(D) $2-\sqrt{2}$

Problem 9.12 Which of the following is NOT equal to $\sqrt{12}$?

(A) $\dfrac{12}{\sqrt{12}}$

(B) $\dfrac{24}{\sqrt{12}}$

(C) $\sqrt{3} \times \sqrt{4}$

(D) $\dfrac{\sqrt{24}}{\sqrt{2}}$

Problem 9.13 Expand $(x+2)(x+4)$ to get $x^2 + Bx + C$. What is B?

Problem 9.14 Expand $(\sqrt{2}+1)(\sqrt{3}+1)$. This equals:

(A) $\sqrt{6}+\sqrt{2}+\sqrt{3}$

(B) $\sqrt{6}+\sqrt{5}+1$

(C) $\sqrt{6}+\sqrt{2}+\sqrt{3}+1$

(D) $\sqrt{5}+\sqrt{2}+\sqrt{3}+1$

Problem 9.15 Which of the following is equal to $(\sqrt{6}+\sqrt{3})(\sqrt{2}-1)$?

(A) $\sqrt{3}$
(B) $\sqrt{12}+\sqrt{6}-\sqrt{3}$
(C) $2\sqrt{6}+\sqrt{3}$
(D) $2\sqrt{3}-\sqrt{6}$

Problem 9.16 The expression $(2\sqrt{7}-\sqrt{5})(2\sqrt{7}+\sqrt{5})$ simplifies to an integer. What is this integer?

Problem 9.17 Write $(\sqrt{11}-1)^2 = A\sqrt{11}+B$ for integers A,B. What is $A+B$?

Problem 9.18 Simplify $\dfrac{4}{\sqrt{7}-\sqrt{3}}$.

(A) $\sqrt{3}-\sqrt{7}$
(B) $4\sqrt{3}-4\sqrt{7}$
(C) $\sqrt{3}+\sqrt{7}$
(D) $4\sqrt{3}+4\sqrt{7}$

Problem 9.19 The equation $\sqrt{x+4}=4$ has one integer solution. What is it?

Problem 9.20 What is the solution to $x\sqrt{2}+x\sqrt{8}=12$?

(A) $x=\sqrt{2}$
(B) $x=2\sqrt{2}$
(C) $x=3\sqrt{2}$
(D) $x=2\sqrt{8}$

9.3 Practice Questions

Problem 9.21 Simplify the following:

(a) $\sqrt{363}$ (write in simplest radical form)

(b) $x^2\sqrt{5}(x+\sqrt{5}+\sqrt{10})$

Problem 9.22 Expand and simplify:

(a) $(4x+3)(x-2)$

(b) $(2x^2+y)(y-3)$

Problem 9.23 Expand and simplify $(\sqrt{7}-3)(\sqrt{14}+\sqrt{2})$

Problem 9.24 Solve the equation $(\sqrt{x}-3)(\sqrt{x}+3)=-2$.

Problem 9.25 Expand and simplify $(\sqrt{2}+\sqrt{6})^2$

Problem 9.26 Simplify the expression $\dfrac{\sqrt{24}}{\sqrt{3}-1}$

Problem 9.27 Simplify $\dfrac{r+\sqrt{r}}{\sqrt{r}-1}$.

Problem 9.28 Solve the equation $1+x=x\sqrt{2}$.

Problem 9.29 Solve $2\sqrt{x-1}-\sqrt{x+2}=0$

Problem 9.30 Expand $(x+\sqrt{2}+1)^2$

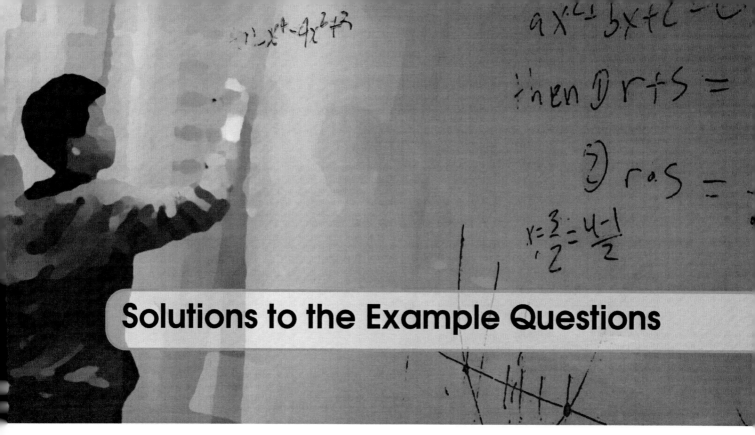

Solutions to the Example Questions

In the sections below you will find solutions to all of the Example Questions contained in this book.

Quick Response and Practice questions are meant to be used for homework, so their answers and solutions are not included. Teachers or math coaches may contact Areteem at info@areteem.org for answer keys and options for purchasing a Teachers' Edition of the course.

1   Solutions to Chapter 1 Examples

Problem 1.1 Evaluate the following expressions.

(a) $3 \times (2 + x)$ if $x = 4$.

Answer

18

Solution

$3 \times (2 + 4) = 3 \times 6 = 18$.

(b) $(k - 1)(k + 2)$ if $k = -1$.

Answer

-2

Solution

$(-1 - 1)(-1 + 2) = (-2)(1) = -2$.

(c) $p^2 + \dfrac{9}{p}$ if $p = 3$.

Answer

12

Solution

$3^2 + \dfrac{9}{3} = 9 + 3 = 12$.

(d) $(a - b)(a + b)$ if $a = 5$ and $b = -3$.

Answer

16

Solution

$(5-(-3))(5+(-3)) = (8)(2) = 16$

(e) $\dfrac{2x+3}{y-x}$ if $x=2$ and $y=6$.

Answer

$\dfrac{7}{4}$

Solution

$\dfrac{2\times 2+3}{6-2} = \dfrac{7}{4}.$

Problem 1.2 For each of the equations below, determine whether the given variables are solutions to the equation.

(a) $-\dfrac{1}{2}x+4 = 2$ if $x=4$.

Answer

Yes.

Solution

The left-hand side is $-\dfrac{1}{2}\times 4+4 = -2+4 = 2$. Since the right-hand side is also 2, $x=4$ is a solution.

(b) $3x-7 = 2x+1$ if $x=5$.

Answer

No.

Solution

The left-hand side is $3\times 5-7 = 15-7 = 8$, while the right-hand side is $2\times 5+1 = 10+1 = 11$. Since $8 \neq 11$, $x=5$ is not a solution.

(c) $3(x-2) = 2(x+2)$ if $x=10$.

Answer

Yes.

Solution

The left-hand side is $3(10-2) = 3(8) = 24$. The right-hand side is $2(10+2) = 2(12) = 24$. Hence $x = 10$ is a solution.

(d) $\dfrac{x+5}{x-5} = 0$ if $x = 5$.

Answer

No.

Solution

Note when we substitute $x = 5$ into the left-hand side we get $\dfrac{5+5}{5-5} = \dfrac{10}{0}$ which is undefined. Therefore $x = 5$ is not a solution.

Problem 1.3 Simplify the following, combining like terms where possible.

(a) $5m - 4 - (m+3)$

Answer

$4m - 7$

Solution

$5m - 4 - (m+3) = 5m - 4 - m - 3 = 4m - 7$.

(b) $0.2(2x+4) - 0.05(5x-4)$

Answer

$0.15x + 1$

Solution

$0.2(2x+4) - 0.05(5x-4) = 0.4x + 0.8 - 0.25x + 0.2 = 0.15x + 1$.

(c) $4 - (6 + 3r - 2s)$

Answer

$-3r + 2s - 2$

Solution

$4 - (6 + 3r - 2s) = 4 - 6 - 3r + 2s = -3r + 2s - 2.$

(d) $2(x + y) + 3(2 - y) + 2$

Answer

$2x - y + 8$

Solution

$2(x + y) + 3(2 - y) + 2 = 2x + 2y + 6 - 3y + 2 = 2x - y + 8$

Problem 1.4 Solve the following equations.

(a) $x - 5 = 10$.

Answer

$x = 15$

Solution

Adding 5 to both sides we get $x - 5 + 5 = 10 + 5$ so $x = 15$.

(b) $y + 3 = 1$.

Answer

$y = -2$

Solution

Subtracting 3 from both sides we get $y + 3 - 3 = 1 - 3$ so $y = -2$.

(c) $3t = 78$

Answer

$t = 26$

Solution

Dividing both sides by 3 we get $\dfrac{3t}{3} = \dfrac{78}{3}$ so $t = 26$.

(d) $4 = \dfrac{z}{6}$

Answer

$z = 24$

Solution

Multiplying both sides by 6 we get $6 \times 4 = 6 \times \frac{z}{6}$ so $24 = z$.

Problem 1.5 Solve the following.

(a) $4 - h = 12$.

Answer

$h = -8$

Solution

First subtract 4 from both sides to get $4 - h - 4 = 12 - 4$ so $-h = 8$. Multiplying both sides by -1 we get $-1 \times -h = -1 \times 8$ so $h = -8$.

(b) $3(x - 3) = 6$

Answer

$x = 5$

Solution 1

Distributing first we have $3x - 9 = 6$. Adding 9 to both sides we get $3x - 9 + 9 = 6 + 9$ so $3x = 15$. Lastly, dividing by 3 we get $\dfrac{3x}{3} = \dfrac{15}{3}$ so $x = 5$.

Solution 2

Alternatively, we can divide by 3 first: $\dfrac{3(x-3)}{3} = \dfrac{6}{3}$ so $x - 3 = 2$. Then adding 3 to both sides we get $x - 3 + 3 = 2 + 3$ so $x = 5$.

Problem 1.6 Solve the following equations.

(a) $\dfrac{4x}{5} = 20$.

Answer

$x = 25$

Solution

Note if we multiply both equations by the reciprocal of $\dfrac{4}{5}$, which is $\dfrac{5}{4}$ the fractions will cancel. This gives

$$\frac{5}{4} \times \frac{4x}{5} = \frac{5}{4} \times 20,$$

so $x = 25$.

(b) $3(c - 0.78) - 2c = 3.57$.

Answer

$c = 5.91$

Solution

Distributing and simplifying, the left-hand side is $3(c - 0.78) - 2c = 3c - 2.34 - 2c = c - 2.34$. Therefore adding 2.34 to both sides we have $c - 2.34 + 2.34 = 3.57 + 2.34$ so $c = 5.91$.

(c) $3(2s - 3) = 4s + 3 - (-2s + 1)$

Answer

No Solution

Solution

Distributing the left-hand side is $6s - 9$. The right-hand side is $4s + 3 + 2s - 1 = 6s + 2$. Therefore we want to solve the equation $6s - 9 = 6s + 2$. Note if we subtract $6s$ from both sides we have $6s - 9 - 6s = 6s + 2 - 6s$ so $-9 = 2$. As this is false, there are no solutions to this equation.

(d) Solve $\dfrac{2}{3}a - 5 = \dfrac{1}{3}a + 2$.

Answer

21

Solution

Adding 5 on both sides we get $\dfrac{2}{3}a - 5 + 5 = \dfrac{1}{3}a + 2 + 5$ so $\frac{2}{3}a = \frac{1}{3}a + 7$. Subtracting $\dfrac{1}{3}a$ from both sides we then isolat a: $\dfrac{2}{3}a - \dfrac{1}{3}a = \dfrac{1}{3}a + 7 - \dfrac{1}{3}a$ so $\dfrac{1}{3}a = 7$. Multiplying by 3 we have $3 \times \dfrac{1}{3}a = 3 \times 7$ so $a = 21$.

Problem 1.7 Solve the following equations for the specified variable.

(a) $3x - 4y = 12$ for x.

Answer

$x = \dfrac{4}{3}y + 4$.

Solution

Adding $4y$ on boths sides we can isolate the $3x$: $3x - 4y + 4y = 12 + 4y$ so $3x = 4y + 12$. Now dividing by 3 we have

$$\frac{3x}{3} = \frac{4y + 12}{3} \text{ so } x = \frac{4y + 12}{3} = \frac{4y}{3} + 4.$$

(b) $x(y + 2) = 3$ for y.

Answer

$$y = \frac{3}{x} - 2$$

Solution

Note if we distribute the x we will get a left-hand side of $xy + 2x$. Subtracting $2x$ on both sides gives $xy + 2x - 2x = 3 - 2x$ so $xy = 3 - 2x$. To isolate the y we divide by x (so we must remember that $x \neq 0$), giving

$$\frac{xy}{x} = \frac{3-2x}{x} \text{ so } y = \frac{3-2x}{x} = \frac{3}{x} - 2$$

(c) $\dfrac{2}{p} = \dfrac{q}{3}$ for p.

Answer

$$p = \frac{9}{q}$$

Solution

To get rid of fractions, we multiply both sides by $3p$. This gives $3p \times \dfrac{3}{p} = 3p \times \dfrac{q}{3}$ so $9 = pq$. (Note, this trick is often referred to as "cross-multiplication".) Thus dividing both sides by q we have $\dfrac{9}{q} = \dfrac{pq}{q}$ so $p = \dfrac{9}{q}$.

Problem 1.8 For each of the following word problems, write and equation and then solve that equation to solve the problem.

(a) George hires a gardener for his house. The gardener charges an initial fee of \$20 plus \$15 per hour that he works. If George pays a total of \$95 to the gardener, how many hours does the gardener work?

Answer

5

Solution

Let h denote the number of hours the gardener works. Then the gardener charges $20 + 15h$ dollars in total for working h hours. This gives us the equation $20 + 15h = 95$.

Subtracting 20 from both sides gives $15h = 75$ and dividing by 15 we have $h = 5$, so the gardener must work 5 hours.

(b) A rectangle is 3 inches longer than it is wide. If the rectangle has a perimeter of 26 inches, what is the area of the rectangle in square inches?

Answer

40

Solution

Let w denote the width of the rectangle, so $w + 3$ is its length. Therefore (as the perimeter is 26 inches) we must have
$$2w + 2(w+3) = 26.$$
Simplifying the left-hand side gives $2w + 2w + 6 = 4w + 6$. Subtracting 6 from both sides we get that $w = 20$ so after dividing by 4 we have $w = 5$. Thus the width of the rectangle is 5 inches and the length is 8 inches. This tells us that the area must be $5 \times 8 = 40$ square inches.

Problem 1.9 Solve the following motion problems. Recall that distance can be calculated using the equation $d = r \times t$ (distance equals rate times time).

(a) William drove to visit his parents. At the start of his trip, there was no traffic, and he drove for 4 hours at his normal speed. For the last 2 hours of the trip, traffic caused William to drive 20 mph less than his normal speed. If William drove 350 miles in total, what is his normal driving speed in mph?

Answer

65

Solution

Let s denote William's normal driving speed, so $s - 20$ is his speed in traffic. In the four hours at normal speed, Williams drives $4 \times s = 4s$ miles. In the two hours in traffic, he drives $2 \times (s - 20) = 2s - 40$ miles. We know that he drives 350 miles in total, so
$$4s + 2s - 40 = 350.$$
Combining like terms on the left-hand side, $6s - 40 = 350$. Adding 40 gives $6s = 390$,

so after dividing by 6 we have $s = \dfrac{390}{6} = 65$. Hence William's normal driving speed is 65 mph.

(b) Parker and Candace run a race. Parker cheats and starts 2 seconds before Candace, but they still end up tying during the race. If Parker can run 6 m/s and Candace can run 6.5 m/s, how long was the race in meters?

Answer

156

Solution

Let t denote the time that Candace spends running. Then Parker runs for $t + 2$ seconds. Therefore we know that Parker runs a distance of $6 \times (t + 2) = 6t + 12$ meters and Candace runs for $6.5 \times t = 6.5t$ meters. Since they tied during the race, we must have $6t + 12 = 6.5t$. Subtracting $6t$ from both sides gives $12 = 0.5t$ so multiplying by 2 on both sides we have $t = 24$. As Candace runs for 24 seconds, the race must have a length of $24 \times 6.5 = 156$ meters.

Problem 1.10 Solve the following problems related to percents.

(a) Laura and Ivy went shopping for clothes during their huge annual sale. Before the discounts, Laura's total was $100 and Ivy's was $80. After the discounts, they remarked that Laura saved $10 more than Ivy. What percentage discount was the store offering during the sale?

Answer

50%

Solution

Suppose the store offers a rate of discount of d (so if the discount is 10%, then $d = 0.10$). Then Laura gets a discount of $100 \times d = 100d$ dollars and Ivy a discount of $80d$ dollars. Since Laura saves 10 dollars more then Ivy, we must have $100d = 80d + 10$. Subtracting $80d$ from both sides we get $20d = 10$, so after dividing by 20 we have that

$$d = \frac{10}{20} = \frac{1}{2} = 0.50.$$

Hence (as $0.5 \times 100 = 50$) the store is offering a 50% discount.

(b) Natasha invested $10,000, split between stocks and bonds. After one year, the stocks had an annual yield of 8% and the bonds had an annual yield of 2%. If Natasha made $560 on the stocks during the year, how many dollars did she invest in bonds?

Answer

$4000.

Solution

Let x denote the amount of money Natasha inversted in bonds. Since she invested 10000 in total, she therefore inversted $10000 - x$ in stocks. Hence in the one year she made

$$x \times 2\% + (10000 - x) \times 8\% = x \times 0.02 + (10000 - x) \times 0.08 = 0.02x + 800 - 0.08x = 800 - 0.06x.$$

Since she made 560 dollars, we know $800 - 0.06x = 560$. Subtracting 800 from both sides we have $-0.06x = -240$ so if we divide by -0.06 we get that

$$x = \frac{-240}{-0.06} = \frac{24000}{6} = 4000.$$

Thus Natasha invested $4000 dollars in bonds.

2 Solutions to Chapter 2 Examples

Problem 2.1 Inequality Warmups: True or False

(a) $-3 \leq 3 < -2$.

Answer

False

Solution

$-3 \leq 3$ is true, but $3 < -2$ is false, so the compound inequality is false.

(b) The inequalities are $-3 < x$ and $x > -3$ are the same.

Answer

True

Solution

Note the order the statements are written does not matter.

(c) The inequalities $x + 2 > 3$ is the same as $2x + 4 > 8$.

Answer

False

Solution

Note dividing the second inequality by 2 on both sides we have $x + 2 > 4$. This is different than $x + 2 > 3$, so the statement is false.

(d) The inequalities $-x > -2$ and $x < 2$ are the same.

Answer

True

Solution

Note if we multiply $-x > -2$ by -1 on both sides, which means we have to flip the $>$ to $<$, we get $x < 2$ so both are the same.

Problem 2.2 Write each of the following inequalities in interval notation and sketch its graph.

(a) $x \geq 3$

Answer

$[3, \infty)$

(b) $x < 10$ and $x > -5$.

Answer

$(-5, 10)$

Solution

$(-\infty, 10) \cap (-5, \infty) = (-5, 10)$.

(c) $x < 2$ or $x > 8$

Answer

$(-\infty, 2) \cup (8, \infty)$

(d) $x < -5$ or $x < 10$

Answer

$(-\infty, 10)$

Solution

$(-\infty, -5) \cup (-\infty, 10) = (-\infty, 10)$ as a number that is less than -5 is automatically less than 10.

Problem 2.3 For each of the inequalities below, determine whether the number satisfies the inequality.

(a) $4x - 5 < 7$ if $x = 2$.

Answer

Yes

Solution

$4x - 5 = 4 \times 2 - 5 = 3$ if $x = 2$. As $3 < 7$ the inequality is true when $x = 2$.

(b) $-4.25x + 32.58 < -1.73$ if $x = 8$.

Answer

No

Solution

If $x = 8$, then $-4.25x + 32.58 = -4.25 \times 8 + 32.58 = -34 + 32.58 = -1.42$. As $-1.42 \not< -1.73$, the inequality is false when $x = 8$.

(c) $-10x + 9 \le 4(x + 2)$ if $x = 1.5$

Answer

Yes

Solution

If $x = 1.5$ the left-hand side is $-10 \times 1.5 + 9 = -15 + 9 = -6$. Note the right-hand side will be positive so since -6 is less than any positive number the inequality is true. (To double check, the right-hand side is $4(1.5 + 2) = 4 \times 4.5 = 18$.

Problem 2.4 Solve the following inequalities for the given variable.

(a) $2x - y > 4$ for y.

Answer

$y < 2x - 4$

Solution

Adding y on both sides we get $2x > y + 4$. Subtracting 4 from both sides we have $2x - 4 > y$ or $y < 2x - 4$.

(b) $\dfrac{2}{3}a - 8b \leq \dfrac{1}{4}$ for a.

Answer

$a \leq 12b + \frac{3}{8}$

Solution

First to get rid of fractions, let's multiply both sides by 12. This gives $8a - 96b \leq 3$. After adding $96b$ to both sides we have $8a \leq 96b + 3$. Dividing by 8 leads to $a \leq 12b + \dfrac{3}{8}$.

Problem 2.5 Humans can roughly hear sounds in the range of $31 - 19,000$ hertz. Cats on the other hand can hear sounds in the range $55 - 77,000$ hertz.

(a) Write the range (in hertz) humans can hear in interval notation. Write the range (in hertz) cats can hear in interval notation.

Answer

Humans: $[31, 19000]$, Cats: $[55, 77000]$

(b) Write the range (in hertz) of sounds that can be heard by both humans and cats in interval notation.

Answer

$[55, 19000]$.

Solution

We want sounds both can hear, so we use intersection: $[31, 19000] \cap [55, 77000] = [55, 19000]$.

(c) Write the range (in hertz) of sounds that can be heard by either humans or cats in interval notation.

Answer

$[31, 77000]$.

Solution

We want sounds both can hear, so we use union: $[31, 19000] \cup [55, 77000] = [31, 77000]$.

Problem 2.6 For each of the following, write and solve an inequality to help answer the question.

(a) A soccer field is 100% longer than it is wide. If the perimeter of the soccer field must be at least 180 meters, what is the range of possible widths for the field?

Answer

The width is at least 30 meters

Solution

Let w be the width of the field. Since the field is 100% longer that it is wide, the field is $w + 100\% \times w = 2w$ meters long. Hence the perimeter is $2w + 2(2w) = 2w + 4w = 6w$. Therefore we must have $6w > 180$. Dividng by 6 we have $w > 30$ so the field is at least 30 meters wide.

(b) Paul and Peter want to buy their father an autographed poster for his birthday. They are looking at posters that range between $80 and $120. Paul is the older brother, and offers to pay $20 more than Peter for the gift. What is the range of money Peter will have to pay for the gift?

Answer

Peter pays between $30 and $50

Solution

If y denotes how many dollars Peter pays for the gift we know that Paul pays $y + 20$ dollars. Therefore we know $80 \le y + y + 20 \le 120$. Hence $80 \le 2y + 20 \le 120$. Subtracting 20 from each we get $60 \le 2y \le 100$. Finally after dividing by 2 we see $30 \le y \le 50$ or in interval notation $(30, 50)$. Hence Peter pays between $30 and $50 for the gift.

(c) Recall that temperatures can be measured in Fahrenheit or in Celsius, where $F = \frac{9}{5}C + 32$ is a formula relating the temperature in Fahrenheit (F) to the temperature in Celsius (C). George does not like the heat, and will not play outside if the temperature is above 95 degrees Fahrenheit. What range of temperatures is too hot for George to play outside measured in Celsius?

Answer

If $C > 35$ it is too hot

Solution

George will not play outside if $F > 95$. Hence he will not play outside if $\frac{9}{5}C + 32 > 95$. Subtracting 32 we have $\frac{9}{5}C > 63$. Multiplying by $\frac{5}{9}$ on both sides we have

$$\frac{5}{9} \times \frac{9}{5}C > \frac{5}{9} \times 63 \text{ or } C > 35.$$

Hence it is too hot for George to play outside if $C > 35$.

Problem 2.7 Solve the following

(a) Harry's grade in his history class is based on the average of 4 tests. On the first three tests he got scores of 45, 75, and 80. Harry wants to get at least a C in the class, so his final grade needs to be at least 70. What is the range of scores Harry can get on the fourth test to ensure he gets at least a C?

Answer

At least 80

Solution

Let x be the grade for the final test. Then Harry's final grade will be

$$\frac{45 + 75 + 80 + x}{4} \text{ so we want } \frac{45 + 75 + 80 + x}{4} \geq 70.$$

Multiplying by 4 on both sides we have $45 + 75 + 80 + x \geq 280$ or $x + 200 \geq 280$. Subtracting 200 from both sides we see $x \geq 80$, so Harry must get at least an 80 on the fourth test.

(b) David and his friends are hungry and order a large pizza for delivery. A pizza costs $11 and each additional topping is $1.50. With tax and tip for the delivery, the final cost is 25% more than the price of the actual pizza. David has at most $20 to spend on the pizza. How many toppings can he and his friends get?

Answer

At most 3 toppings

Solution

Let t be the number of toppings on the pizza. Then the cost of the pizza is $11 + 1.5t$ dollars. With a final cost of 25% more or 125% of the original price, the total price they pay for the pizza is

$$125\% \times (11 + 1.5t) = 1.25 \times (11 + 1.5t)$$

dollars. Hence we must have $1.25 \times (11 + 1.5t) \leq 20$. Dividing by 1.25 first on both sides, we have $11 + 1.5t \leq 16$. Subtracting 11 we get $1.5t \leq 5$. Finally dividing by 1.5 we have $t \leq \dfrac{5}{1.5} = \dfrac{10}{3} \approx 3.33$. However, fractional toppings are not possible, so David and his friends can get at most 3 toppings.

Problem 2.8 How many different sets of four consecutive positive even integers whose sum is less than 100 are there in total?

Answer

10

Solution

Let x be the first positive even integer. Then we know $x \geq 2$. As the other three numbers are $x+2, x+4, x+6$, we also know that $x+x+2+x+4+x+6 < 100$. Working with this second inequality, if we collect like terms we have $4x + 12 < 100$. Subtracting 12 on both sides gives $4x < 88$ or (after dividing by 4) $x < 22$.

Thus $x \geq 2$ and $x < 22$ or x is in the interval $[2, 22)$. Since x must be an even integer, we know x can be $2, 4, 6, 8, \ldots, 20$. This is a total of 10 different values of x. As each of these leads to a different set of 4 consecutive positive even integers, the answer is 10.

Problem 2.9 Solve the inequality $x^2 + 4 \geq 8$.

Answer

$x \geq 2$ or $x \leq -2$: $(-\infty, -2) \cup (2, \infty)$.

Solution

Subtracting 4 from both sides we get $x^2 \geq 4$. Note that $2^2 = (-2)^2 = 4$. Therefore we see if $x \geq 2$ then $x^2 \geq 4$. Similarly if $x \leq -2$, then $x^2 \geq 4$. As both of these work, we must use an 'or' statement (a union) to get our final answer of $(-\infty, -2) \cup (2, \infty)$.

Problem 2.10 Find all z such that $10 - z^2 > -71$ and $5z - 3 > 2(z+2)$.

Answer

$\left(\frac{7}{3}, 9\right)$

Solution

For the first inequality, if we add 71 to both sides we have $81 - z^2 > 0$ or after adding z^2 to both sides, $81 > z^2$. For $z^2 < 81$ we see that $-9 < z < 9$ or z is in the interval $(-9, 9)$.

For the second inequality, distributing gives us $5z - 3 > 2z + 4$. Adding 3 to both sides and then subtracting $2z$ from both sides gives $3z > 7$. Hence after dividing by 3 we get $z > \dfrac{7}{3}$ or $\left(\dfrac{7}{3}, \infty\right)$.

Hence we want z in the interval $(-9, 9) \cap \left(\dfrac{7}{3}, \infty\right) = \left(\dfrac{7}{3}, 9\right)$ or $\dfrac{7}{3} < z < 9$.

3 Solutions to Chapter 3 Examples

Problem 3.1 Find the value of the following absolute value expressions.

(a) $|-3| + |4|$

Answer

7

Solution

$|-3| + |4| = 3 + 4 = 7$

(b) $-|3| - |-5|$

Answer

-8

Solution

$-|3| - |-5| = -3 - 5 = -8$

(c) $|2||-4+8| - |-4||5-3|$

Answer

0

Solution

$|2||-4+8| - |-4||5-3| = (2)(4) - (4)(2) = 8 - 8 = 0$

Problem 3.2 Plug in the variables in the expressions.

(a) $|2x - 4| + |-x|, x = 6$

Answer

14

Solution

$|2(6) - 4| + |-(6)| = |8| + |-6| = 8 + 6 = 14$

(b) $3|-2x + 1| - |x + 8|, x = -5$

Answer

24

Solution

$3|-2(-5) + 1| - |(-5) + 8| = 3|-9| - |3| = 3(9) - 3 = 24$

Problem 3.3 Solve the following absolute value equations.

(a) $|x| = 10$

Answer

$x = -10, 10$

Solution

If $x \geq 0$, then $|x| = x$ and so $x = 10$. If $x < 0$, then $|x| = -x$ and so $-x = 10$. Therefore the two possible solutions are $x = 10$ and $x = -10$.

(b) $|-3 + x| = 8$

Answer

$x = -5, 11$

Solution

If $-3 + x \geq 0$, then $-3 + x = 8$, so $x = 11$. If $-3 + x < 0$, then $-(-3 + x) = 8$, so $x = -5$. Therefore the two possible solutions are $x = 11$ and $x = -5$.

(c) $|-8x + 4| = -3$

Answer

No solution

Solution

Absolute values are always nonnegative. This means that there is no way that the absolute value on the left of the equation is equal to the negative number on the right.

Problem 3.4 Solve the following absolute value inequalities and graph the solutions on the number line.

(a) $|x| \leq 3$

Answer

$-3 \geq x \geq 3$

Solution

If $x < 0$, then $-x \leq 3$, so $x \geq -3$. If $x \geq 0$, then $x \leq 3$. Therefore $-3 \leq x < 0$ or $0 \leq x \leq 3$, that is, $-3 \geq x \geq 3$.

(b) $|x| > 2$

Answer

$(-\infty, -2) \cup (2, \infty)$

Solution

If $x \geq 0$, then any x such that $x > 2$ will be a solution. If $x < 0$, then any number x such that $x < -2$ will be a solution. Thus, the solutions are $x < -2$ and $x > 2$. Note that these are the same solutions we would get for the inequality $x^2 > 4$.

(c) $|x+2| - 1 \leq 3$

Answer

$-6 \leq x \leq 2$: $[-6, 2]$.

Solution

Adding 1 to both sides we have $|x+2| \leq 4$. Note $x+2 \geq 0$ if $x \geq -2$, so we break into the cases of $x \geq -2$ and $x < -2$.

If $x \geq -2$ (so $x + 2 \geq 0$), $|x+2| = x+2$ so we want $x + 2 \leq 4$ or $x \leq 2$.

If $x < -2$ (so $x + 2 < 0$), $|x+2| = -(x+2) = -x-2$ so we want $-x - 2 \leq 4$ or $-x \leq 6$ or $x \geq 6$.

Combining we have $-6 \leq x \leq 2$ or $[-6, 2]$.

Problem 3.5 Solve the following absolute value inequalities.

(a) $3|x-2| + 3 < 12$

Answer

$-1 < x < 5$: $(-1, 5)$

Solution

Subtracting 3 and then dividing by 3 gives the inequality $|x-2| < 3$. We use the cases $x \geq 2$ (where $x - 2 \geq 0$) and $x < 2$ (where $x - 2 < 0$).

If $x \geq 2$ then $|x-2| = x - 2$ so we need $x - 2 < 3$ or $x < 5$. If $x < 2$ then $|x-2| = -(x-2) = -x+2$ so we need $-x + 2 < 3$ or $-x < 1$ so $x > -1$.

Combining we get an answer of $-1 < x < 5$ or $(-1, 5)$.

(b) $|x+4| + 4 < 12$

Answer

$-16 < x < 8$: $(-16, 8)$

Solution

First subtracting 4 gives $|x+4| < 8$.

If $x + 4 \geq 0$, that is, if $x \geq -4$, then $x + 4 < 12$, so $-4 \leq x < 8$. If $x + 4 < 0$, that is, if $x < -4$, then $-(x+4) < 12$, that is $-x < 16$, so $-16 < x < -4$. Therefore $-16 < x < 8$.

(c) $2|3 - 2x| - 6 \geq 18$

Answer

$x \le -\frac{9}{2}$ or $\frac{15}{2} \le x$: $\left(-\infty, -\frac{9}{2}\right] \cup \left(\frac{15}{2}, \infty\right)$

Solution

Adding 6 and then dividing by 2 gives the inequality $|3 - 2x| \ge 12$.

If $3 - 2x \ge 0$, that is, if $x \le \dfrac{3}{2}$, then $3 - 2x \ge 12$. Subtracting 3 gives $-2x \ge 9$ so after dividing by -2 (and flipping the inequality) we have $x \le \dfrac{-9}{2}$.

If $3 - 2x < 0$, that is, if $x > \dfrac{3}{2}$, then $-(3 - 2x) = 2x - 3 \ge 12$. Adding 3 gives $2x \ge 15$, so $x \ge \dfrac{15}{2}$.

Hence $x \le \dfrac{-9}{2}$ or $x \ge \dfrac{15}{2}$.

Problem 3.6 Solve the following absolute value equations.

(a) $|x| + 3 = |2x|$

Answer

$x = -3, 3$

Solution

Note if $x \ge 0$ then $|x| = x$ and $|2x| = 2x$ so we have the equation $x + 3 = 2x$ so $x = 3$. If $x < 0$ then $|x| = -x$ and $|2x| = -2x$ so we have the equation $-x + 3 = -2x$ so $3 = -x$ or $x = -3$. Hence the solutions are $x = -3, 3$.

(b) $|x + 3| = |x|$

Answer

$x = \frac{-3}{2}$

Solution

If $x \geq 0$, then $|x| = x$ and $|x+3| = x+3$ so we have the equation $x+3 = x$ or $3 = 0$ which is impossible.

If $x < 0$ then $|x| = -x$, but since $x+3$ could still be positive or negative, we need to break the problem into further cases. If $x \geq -3$, then $x+3 \geq 0$ so $|x+3| = x+3$. This gives the equation $x+3 = -x$ or $2x = -3$ so $x = -\frac{3}{2}$. Lastly, if $x < 3$ we have $|x+3| = -x-3$ giving the equation $-x-3 = -x$ or $-3 = 0$ so again there are no solutions.

Hence the only solution is $x = -\frac{3}{2}$.

Problem 3.7 Solve the following equations with nested absolute values.

(a) $|3 + |x+2|| = 4$

Answer

$x = -1, -3$

Solution

First note we must have $3 + |x+2| = 4$ or $3 + |x+2| = -4$. Since absolute values are always positive the second case is impossible so we know that $3 + |x+2| = 4$ or $|x+2| = 1$. If $x+2 = 1$ then $x = -1$ while if $x+2 = -1$ we have $x = -3$. Thus our two answers are $x = -1$ and $x = -3$.

(b) $|x - |2x+1|| = 3$

Answer

$x = -\frac{4}{3}, 2$

Solution

We know that $|2x+1| = 2x+1$ if $x \geq -\frac{1}{2}$ and $|2x+1| = -2x-1$ if $x < -0.5$.

First assume $x \geq -\frac{1}{2}$ so we have $|x - 2x - 1| = 3$ or $|-x-1| = 3$. This is the same as $|x+1| = 3$. If $x+1 = 3$ we have $x = 2$ as a solution and if $x+1 = -3$ we get $x = -4$.

However, remember we are assuming $x \geq -\dfrac{1}{2}$ for this case, so $x = -4$ does not work. Hence $x = 2$ is the only solution with $x \geq -\dfrac{1}{2}$.

If $x < -\dfrac{1}{2}$ we have $|x + 2x + 1| = 3$ or $|3x + 1| = 3$. If $3x + 1 = 3$ we get $x = \dfrac{2}{3}$. However, this is not less than $-\dfrac{1}{2}$, so it does not work. Else we have $3x + 1 = -3$ or $x = -\dfrac{4}{3}$.

Therefore $x = -\dfrac{4}{3}$ and $x = 2$ are the solutions.

Problem 3.8 Find all values of x, y and z such that $|x - 6| + |y^2 - 4| + |z + 8| = 0$

Answer

$x = 6, y = -2, z = -8$ or $x = 6, y = 2, z = -8$

Solution

Absolute values are always nonnegative, thus each of the absolute values in the equation must be equal to 0. This tells us that $|x - 6| = 0$, $|y^2 - 4| = 0$ and $|z + 8| = 0$, that is, $x = 6$, $y = -2, 2$ and $z = -8$.

Problem 3.9 Solve the following absolute value equations.

(a) $|x + 1| + |3(x + 1)| = 8$

Answer

$x = -3, 1$

Solution

Note we can rewrite the equation as $|x + 1| + 3|x + 1| = 9$ or $4|x + 1| = 8$. Hence we must have $|x + 1| = 2$. Thus $x + 1 = 2$ implying $x = 1$ or $x + 1 = -2$ implying $x = -3$.

(b) $4|x - 6| - |7x - 42| \geq -3$

Solutions to the Example Questions

Answer

$5 \leq x \leq 7$

Solution

Rewriting the equation we have $4|x-6| - 7|x-6| \geq -3$ or $-3|x-6| \geq -3$. Dividing by -3 and flipping the inequality we have $|x-6| \leq 1$. Thus $-1 \leq x-6 \leq 1$ or $5 \leq x \leq 7$.

Problem 3.10 Write an absolute value inequality that is equivalent to the compound inequality.

(a) $-3 \leq x \leq 3$

Answer

$|x| \leq 3$

(b) $-2 \leq x \leq 4$

Answer

$|x-1| \leq 3$

Solution

Note if we subtract one from the inequality we have $-3 \leq x-1 \leq 3$. This can be rewritten as $|x-1| \leq 3$ for our answer.

(c) $\frac{3}{2} < x < \frac{9}{2}$

Answer

$|2x-6| < 3$

Solution

First let's get rid of fractions by multiplying by 2: $3 < 2x < 9$. Now by subtracting 6 from both sides we have $-3 < 2x-6 < 3$ or $|2x-6| < 3$.

Copyright © ARETEEM INSTITUTE. All rights reserved.

4   Solutions to Chapter 4 Examples

Problem 4.1 On a single coordinate plane, plot the following points: $A = (0,1)$, $B = (2,-3)$, $C = (5,1)$, $D = (-1,5)$, $E = (-3,-3)$. For each of the points, state explain its position in the plane (which axis it is on, which quadrant it is in, etc.).

Answer

A: y-axis, B: Quadrant IV, C: Quadrant I, D: Quadrant II, E: Quadrant III

Problem 4.2 Graph the following equations on a coordinate plane.

(a) $y = 2x - 1$.

Answer

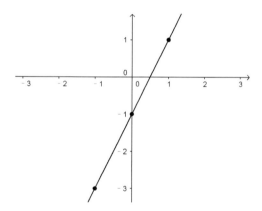

Solution

Plotting a few points we get $(-1,-3)$, $(0,-1)$, and $(1,1)$ which when connected produce the graph of the line as needed.

(b) $y = x^2$.

Answer

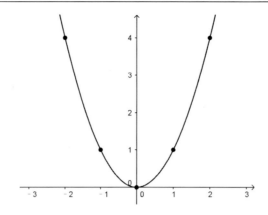

Solution

Plotting a few points we get $(-2,4)$, $(-1,1)$, $(0,0)$, $(1,1)$, and $(2,4)$ which when connected produce the graph as needed.

(c) $y = |x|$

Answer

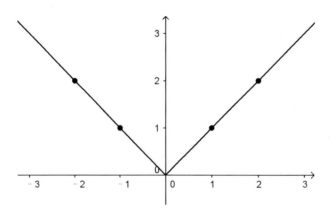

Solution

Plotting a few points we get $(-2,2)$, $(-1,1)$, $(0,0)$, $(1,1)$, and $(2,2)$ which when connected produce the graph as needed.

Problem 4.3 Find the slope between the two given points.

(a) $(2,3)$ and $(4,9)$.

Answer

3

Solution

The slope is

$$\frac{9-3}{4-2} = \frac{6}{2} = 3.$$

(b) $(-2,4)$ and $(6,4)$.

Answer

0

Solution

The slope is

$$\frac{4-4}{6-(-2)} = \frac{0}{8} = 0.$$

Note there is no change in y, so we can see the slope must be 0.

(c) $(-3,5)$ and $(1,0)$.

Answer

$-\frac{5}{4} = -1.25$

Solution

The slope is

$$\frac{0-5}{1-(-3)} = \frac{-5}{4} = -\frac{5}{4}.$$

(d) $(2,0)$ and $(2,-1)$.

Answer

Undefined.

Solution

The slope is

$$\frac{-1-0}{2-2} = \frac{-1}{0} = \text{undefined}$$

as we cannot divide by 0. Alternatively, we can see there is no change in x, so the slope must be undefined.

Problem 4.4 For each of the following relations, classify the following as (i) a linear function, (ii) a function that is not linear, or (iii) not a function.

(a)

x	1	2	3	5	8
$f(x)$	2	3	4	6	9

Answer

Linear

Solution

We can see that whenever x increases by 1, $f(x)$ increases by 1, so the function is linear. In fact, the equation for $f(x)$ can be given by $f(x) = x + 1$.

(b)

x	-2	-1	0	1	4
$f(x)$	-4	-1	0	-1	-16

Answer

Function but not linear

Solution

This is a function, as for every input there is only one output. However, because there are repeated values for the output $f(-1) = f(1) = -1$ but the function is not constant, it cannot be linear. In fact, the equation for $f(x)$ can be given by $f(x) = -x^2$.

(c)

x	1	1	2	2	4	4
$f(x)$	-1	1	-2	2	-4	4

Answer

Not a function

Solution

Note that 1 and -1 are outputs for the input 1, so this is not a function.

Problem 4.5 Write a linear equation in slope-intercept form for each of the following.

(a) A linear equation with slope -2 containing the point $(0,2)$.

Answer

$y = -2x + 2$

Solution

The point $(0,2)$ means that the y-intercept is 2, so $b = 2$. We are given that the slope $m = -2$, so the line has equation $y = -2x + 2$.

(b) A linear equation with slope 4 containing the point $(-2,7)$.

Answer

$y = 4x + 1$

Solution

We are given the slope and a point on the line, so using point-slope form we have $y - 7 = 4(x - (-2))$ is the equation for the line. Rewriting in slope-intercept form we have $y - 7 = 4x + 8$ so $y = 4x + 1$ is our answer.

Problem 4.6 Write a linear equation in point-slope form for each of the following.

(a) A linear equation with slope -2 containing the point $(3,7)$.

Answer

$y - 7 = -2(x - 3)$

Solution

We know the slope and a point on the line, so we can directly use point-slope form to have our answer of $y - 7 = -2(x - 3)$.

(b) A linear equation containing the points $(2,3)$ and $(8,2)$.

Answer

$y - 3 = -\frac{1}{6}(x-2)$ or $y - 2 = -\frac{1}{6}(x-8)$

Solution

We are given two points, so we first find the slope. The slope is

$$\frac{2-3}{8-2} = \frac{-1}{6} = -\frac{1}{6}.$$

Hence using point-slope form we have an answer of

$$y - 3 = -\frac{1}{6}(x-2) \text{ or } y - 2 = -\frac{1}{6}(x-8).$$

Note if you simplified both of these equations you would see they are the same!

Problem 4.7 Write a linear equation in standard form for each of the following.

(a) The linear equation $y = 2x + 4$.

Answer

$2x - y = -4$.

(b) The linear equation with x-intercept 2 and y-intercept 4.

Answer

$2x + y = 4$.

Solution

Since 4 is a multiple of 2 and 4 let's start by assuming $C = 4$ (we can change this later if needed). Then the equation will have the form $Ax + By = 4$. Since the x-intercept is 2, we know that $A \times 2 + B \times 0 = 4$, so $A = 2$. Similarly as the y-intercept is 4, we know that $A \times 0 + B \times 4 = 4$ so $B = 1$. Hence the line is $2x + y = 4$.

Problem 4.8 Graph the following linear inequalities.

(a) $y > -\frac{1}{2}x + 2$.

Answer

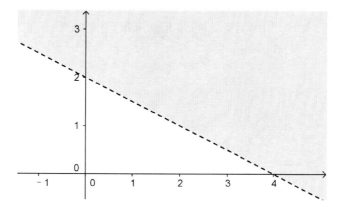

Solution

The line has a y-intercept of 2 and slope of $-\dfrac{1}{2}$. Since we want y greater than this line, but not including the line, we shade the region above and use a dotted line for the graph.

(b) $2x - 3y < 12$

Answer

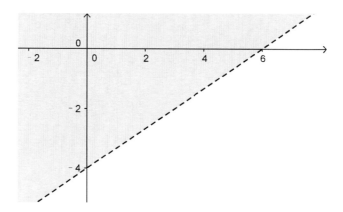

Solution

We can calculate the x-intercept as $12 \div 2 = 6$ and similarly the y-intercept as $12 \div -3 = -4$. This produces the line as above. Checking the point $(0,0)$ we see that

$2 \times 0 - 3 \times 0 < 12$, so we want to shade the half containing $(0,0)$ which produces the graph above.

Problem 4.9 Graph the following equations with absolute values.

(a) $y = |x|$.

Answer

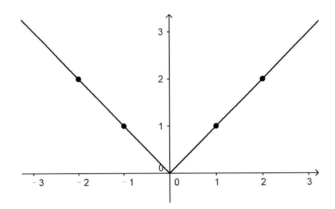

Solution

Note we have already graphed this by plotting points. More directly, note that $|x| = x$ when $x \geq 0$ and $|x| = -x$ when $x < 0$. Hence we need to plot the line $y = x$ when $x \geq 0$ and $y = -x$ when $x < 0$ which produces the same graph.

(b) $y = 2|x-2|$.

Answer

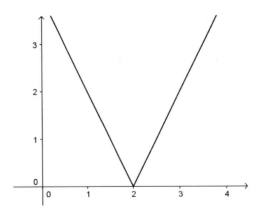

Solution

For when $x \geq 2$ we have $|x-2| = x-2$ so $y = 2(x-2)$ or $y = 2x-4$. Similarly when $x < 2$ we have $|x-2| = -x+2$ so $y = 2(-x+2)$ so $y = -2x+4$. Putting the portions of these lines together gives the graph above.

(c) $|2x - y| = 1$.

Answer

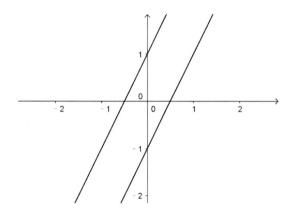

Solution

Note we either have $2x - y = 1$ so $y = 2x - 1$ or $2x - y = -1$ so $y = 2x + 1$. Both lines have slope 2, but y-intercepts of either 1 or -1. Graphing them together gives the result above.

Problem 4.10 Graph the following inequalities with absolute values.

(a) $y > |x+1|$.

Answer

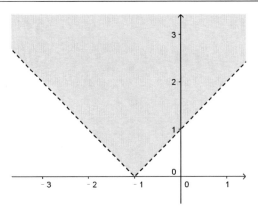

Solution

First note we want the region strictly above $y = |x+1|$. When $x \geq -1$ we have $|x+1| = x+1$ so we have the line $y = x+1$. Otherwise when $x < -1$ we get $|x+1| = -x-1$ so $y = -x-1$. Combining the portions and shaded the region above gives the graph.

(b) $|x+y| < 2$.

Answer

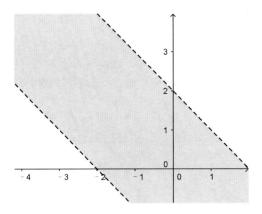

Solution

Note we must have that $x+y < 2$ which is $y < -x+2$ and that $-(x+y) < 2$ which is $x+y > -2$ or $y > -x-2$. Combining these two regions gives the graph above.

5 Solutions to Chapter 5 Examples

Problem 5.1 Find the equations for the following lines.

(a) The line with x-intercept 4 and slope -2.

Answer

$y = -2x + 8$

Solution

We know the line contains the point $(4, 0)$ and has slope -2. Using point-slope form we have the line is

$$y - 0 = -2(x - 4) \text{ which simplifies to } y = -2x + 8.$$

(b) The line containing the points $(-2, -4)$ and $(2, 6)$.

Answer

$y = \frac{5}{2}x + 1$

Solution

We first calculate the slope:
$$\frac{6 - (-4)}{2 - (-2)} = \frac{10}{4} = \frac{5}{2}.$$
Then using point-slope form we have
$$y - 6 = \frac{5}{2}(x - 2) \text{ which simplifies to } y = \frac{5}{2}x + 1.$$

Problem 5.2 For each of the following pairs of lines determine whether they are parallel, perpendicular, or neither.

(a) $y = 2x + 3$ and $y = -\frac{1}{2}x + 2$

Answer

Perpendicular

Solution

We see the slopes are opposite reciprocals so the lines are perpendicular.

(b) $y = -\frac{3}{4}x + 4$ and $3x + 4y = 24$.

Answer

Parallel

Solution

Rewriting the second equation we have $4y = -3x + 24$ or $y = -\dfrac{3}{4}x + 6$ so the lines are parallel.

(c) The line containing the points $(2, 0)$ and $(-1, 3)$ and the line $y = x - 4$.

Answer

Perpendicular

Solution

Note the slope of the first line is

$$\frac{3 - 0}{-1 - 2} = \frac{3}{-3} = -1.$$

Since the slope of the second line is 1 they are perpendicular lines.

Problem 5.3 Find equations for the lines described below.

(a) The line containing the point $(0, 2)$ that is parallel to the line containing the points $(1, 2)$ and $(3, 4)$.

Answer

$y = x + 2$

Solution

We know that the line has y-intercept 2, so we just need to find the slope. The slope between the points $(1,2)$ and $(3,4)$ is

$$\frac{4-2}{3-1} = \frac{2}{2} = 1$$

so since our line is parallel it has the same slope. Hence the line we want has equation $y = x + 2$.

(b) The line containing the origin that is perpendicular to the line $2x + y = 4$.

Answer

$y = \dfrac{1}{2}x$

Solution

The line $2x + y = 4$ can be rewritten as $y = -2x + 4$, so it has slope -2. Thus the line we want has slope $\dfrac{1}{-2} = -\dfrac{1}{2}$ and hence equation $y = -\dfrac{1}{2}x$ as it contains the origin.

Problem 5.4 Write equations describing the following situations. If they describe a direct varation, inverse varation, or joint variation, say which variation is described.

(a) Larry has \$400 to spend on collectible action figures. If the price of each action figure is P, write an equation describing N, the number of action figures Larry can buy.

Answer

$N = \dfrac{400}{P}$; Inverse Variation

Solution

If Larry buys N action figures that cost P each we know he spends $N \times P$ dollars in total. Hence $N \times P = 400$ so N varies inversely with P and if we solve for N we get $N = \dfrac{400}{P}$.

(b) William gets paid \$8 an hour for his work. If he works for T hours, write an equation describing P, the total payment William gets for his work.

Answer

$P = 8T$; Direct Variation

Solution

The total amount of money William is paid is equation to $8 \times T$ as he earns $8 per hour. Hence $P = 8T$ and P varies directly with T.

(c) Cary drives a semi and gets paid $0.75 per mile he drives. If he drives for T hours at R miles per hour, write an equation describing P, the total payment Cary gets for driving.

Answer

$P = 0.75RT$; Joint Variation

Solution

Cary gets paid $0.75 per mile, so his payment varies directly with the distance he travels. However, we know the distance is equal to $R \times T$, so $P = 0.75RT$ and P varies jointly with R and T.

(d) Patricia sells printers at the local office supplies store. He gets paid a base salary of $200 per week, and earns 4% commission on her sales. If she makes a total of S dollars in sales this week, write an equation describing W, Patricia's total wage for the week.

Answer

$W = 200 + 0.04S$; None of the described variations

Solution

Patricia's commission is 4% of her total sales, so is $4\% \times S = 0.04S$. She also gets a base salary of $200 so we have $W = 200 + 0.04S$. While this is a linear relationship, it is not a direct, inverse, or joint variation. (It is true, however, that her commission varies directly with her sales.)

Problem 5.5 Fill in the missing values in the tables below.

x	y
0.5	
1	300
	10
600	

(a) where y is inversely proportional to x.

Missing Pairs: $(0.5, 600)$, $(30, 10)$, $(600, 0.5)$

Solution

y is inversely proportional to x, so $x \times y = k$ for some constant k. We see that $K = 1 \times 300 = 300$. Hence if $x = 0.5$ then $y = 600$, if $y = 10$, then $x = 30$, and finally if $x = 600$ then $y = 0.5$.

x	y
$\frac{1}{3}$	$\frac{1}{4}$
	6
12	
20	

(b) where y is directly proportional to x.

Answer

$(8, 6)$, $(12, 9)$, $(20, 15)$

Solution

y is directly proportional to x, so $y = k \times x$ for some constant k. We know that

$$\frac{1}{4} = \frac{1}{3} \times k \Rightarrow k = \frac{1}{4} \times 3 = \frac{3}{4}.$$

Hence when $y = 6$, $x = \dfrac{4}{3} \times 6 = 8$, when $x = 12$, $y = \dfrac{3}{4} \times 12 = 9$, and when $x = 20$, $y = \dfrac{3}{4} \times 20 = 15$.

Problem 5.6 A football is thrown downward from the top of a building. Its velocity is 38 feet per second after 1 second and 70 feet per second after 2 seconds.

(a) The velocity v of the football can be given using a linear equation in terms of the time t. What is this equation?

Answer

$v = 32t + 6$.

Solution

Note the velocity changes by $70 - 38 = 32$ feet per second in $2 - 1 = 1$ seconds, so the slope of the line is 32. If the velocity increases by 32 each second, we see that the initial velocity is $38 - 32 = 6$. Hence the line is $v = 32t + 6$.

(b) What was the initial velocity of the ball?

Answer

6 ft/sec

Solution

Note the initial velocity is the intercept of 6 from part (a), so the initial velocity is 6 ft/sec.

(c) What is the velocity of the football after 4.5 seconds?

Answer

150 ft/sec

Solution

Using the equation from part (a), we have that the velocity after 4.5 seconds (when $t = 4.5$) is $32 \times 4.5 + 6 = 144 + 6 = 150$ ft/sec.

(d) After how many seconds is the ball traveling 100 feet per second? Round your answer to the nearest tenth if necessary.

Answer

2.9

Solution

We want to solve for t when $v = 100$. Hence $100 = 32t + 6$ or $32t = 94$ so

$$t = \frac{94}{32} = \frac{47}{16} = 2 + \frac{5}{16} \approx 2.9.$$

Problem 5.7 William wants to enclose a rectangular portion of his backyard using a fence. For the East and West sides of his enclosure he will use one type of fence that costs $8 per yard. For the other two sides (North and South) he uses a fence that costs $5 per yard. William wants to spend at most $200 on his fence.

(a) Write an inequality in terms of the width W (the length of the East and West sides) and the length L (the length of the North and South sides) to help William make sure the size he chooses fits his budget.

Answer

$16W + 10L \leq 200$.

Solution

Note the total amount of fence on the East and West sides that William will use is $2W$, so its cost will be $\$8 \times 2W = 16W$ dollars. Similarly the cost on the North and South sides combined will be $\$5 \times 2L = 10L$ dollars. Since he wants to spend at most $200, we must have $16W + 10L \leq 200$.

(b) William ends up deciding that he wants the length L to be twice the width W. If the dimensions of the rectangular enclosure are integers, what is the area of the largest enclosure William can build?

Answer

50 square yards.

Solution

If we know that the length L is twice the width we have $L = 2W$ so we can rewrite our inequality as

$$16W + 10(2W) \leq 200 \text{ so } 36W \leq 200 \text{ so } W \leq \frac{200}{36} = \frac{50}{9}.$$

Since W must be an integer and $\frac{50}{9} = 5 + \frac{5}{9}$, we know that W must be 5 yards. Therefore the length is 10 yards and hence the ensclosure has area $5 \times 10 = 50$ square yards.

Problem 5.8 Consider the data given below.

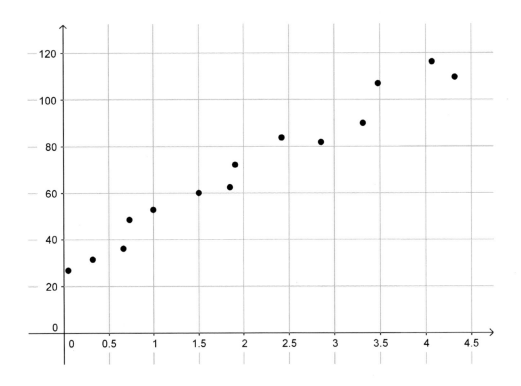

Model the data using a line that has a slope and y-intercept that are both multiples of 10.

Answer

$y = 20x + 30$.

Solution

Drawing an approximate line we get something very similar to:

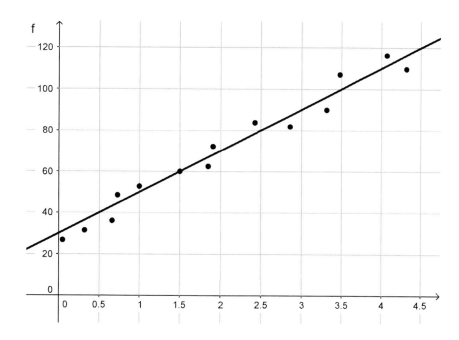

From this line we can see that the y-intercept is roughly 30 and that the line contains the point $(4.5, 120)$. Hence the line has slope

$$\frac{120 - 30}{4.5 - 0} = \frac{90}{4.5} = 20.$$

Therefore the model line is $y = 20x + 30$.

Problem 5.9 Piper has a pepper farm. Each summer, before growing season, she uses some horses to help prepare the field. Last year she used 4 horses and it took her 75 minutes. This year she used 6 horses and it took her 50 minutes. Next year she plans to use 8 horses.

(a) Suppose the time it takes is inversely proportional to the number of horses used. How many minutes will it take Piper to prepare the pepper field next year?

Answer

37.5 minutes.

Solution

We know that it takes 75 minutes for Piper to prepare the pepper field with 4 horses. If the time is inversely proportional to the number of horses, doubling the horses from 4 to 8 will halve the time from 75 to 37.5 minutes.

(b) Suppose the time it takes follows a linear equation based on the number of horses used. How many minutes will it take Piper to prepare the pepper field next year?

Answer

25 minutes.

Solution

We know it takes 75 minutes to prepare the field with 4 horses and 50 minutes with 6 horses. Hence if the time it takes follows a linear relationship with the number of horses, we see that adding 2 horses reduces the time by 25 minutes, so increasing the horses from 6 to 8 decreases the time from 50 to $50 - 25 = 25$ minutes.

(c) Which of the models (described in parts (a) and (b)) do you think is more accurate? Explain your answer.

Solution

The inversely proportional model is more accurate. More horses help the work go faster, but at a decreasing rate. (It may be hard for Piper to control all the horses!) Further, note that with the linear model decreasing the horses to 10 means Piper can prepare the field in 0 time, which does not make sense.

Problem 5.10 Lauren studies the growth of plants at a laboratory. Lauren hypothesizes that the height of the plants she studies is directly proportional to the number of days of growth with a growth rate of 1.5 inches per day.

(a) Write an equation describing Lauren's hypothesis. Use H to represent the plant's height in inches and D to represent the number of days.

Answer

$H = 1.5D$.

Solution

The growth rate of 1.5 inches per day is the constant of proportion, so $H = 1.5D$ in a direct variation.

(b) Before looking at her data, Lauren decides that she will accept her hypothesis as long as the height of the plants is within 1 inch (inclusive) of her model. Write an inequality using absolute values that the data must satisfy for Lauren to accept her hypothesis.

Answer

$|H - 1.5D| \leq 1$

Solution

Note H is the height of the plant, which is estimated by $1.5D$ by Lauren's model. Since we want to the height to be within 1 inch of the prediction (and we don't care whether the height is above or below the predicted value) we must have $|H - 1.5D| \leq 1$.

6   Solutions to Chapter 6 Examples

Problem 6.1 Graph the following pairs of lines on one coordinate plane. Do they intersect? If so, where?

(a) (i) $y = 2x - 3$, (ii) $y = 4x + 3$

Answer

Yes, at $(-3, -9)$

Solution

When the graphs intersect, they do so when their y values are the same, so we must have $2x - 3 = 4x + 3$. Collecting like terms we get $2x = -6$ so $x = -3$. Substituting we get $y = 2(-3) - 3 = -9$, so they lines intersect at the point $(-3, -9)$.

(b) (i) $2x + y = 3$, (ii) $4x + 2y = -4$

Answer

No

Solution

Rewriting in slope-intercept form we get $y = -2x + 3$ for (i) and $y = -2x - 2$ for (ii). We see these lines are parallel (since they have the same slope) so they do not intersect. Double checking this, we must have $-2x + 3 = -2x - 2$ or $3 = -2$ so this is impossible as claimed.

(c) (i) $x - 2y = 4$, (ii) $y + 1 = \dfrac{1}{2}(x - 2)$

Answer

Yes, whenever $y = \dfrac{1}{2}x - 2$

Solution

Rewriting the equations in slope-intercept form we have $-2y = -x + 4$ or $y = \dfrac{1}{2}x - 2$ for (i) and for (ii) we have $y = -1 + \dfrac{1}{2}x - 1$ which is $y = \dfrac{1}{2}x - 2$. Note these equations are the same, so the lines intersect infinitely many times at any point on this line.

Problem 6.2 Solve the following systems of equations using substitution.

(a) (i) $y = 2x - 3$, (ii) $2y - x = 3$

Answer

$(3,3)$

Solution

Subsituting the first equation into the second we get $2(2x - 3) - x = 3$ so $4x - 6 - x = 3$ and hence $3x = 9$. Thus $x = 3$ and $y = 2(3) - 3 = 3$.

(b) (i) $x - y = 3$, (ii) $3x - 2y = 4$

Answer

$(-2, -5)$

Solution

Solving the first equation for x we have $x = y + 3$. Substiting this into the second equation we get $3(y + 3) - 2y = 4$ which is $3y + 9 - 2y = 4$ after distributing and $y = -5$ after combining like terms. Hence $x - (-5) = 3$ so $x = -2$ and $(-2, -5)$ is the solution.

Problem 6.3 Solve the following systems using elimination.

(a) (i) $x + y = 10$, (ii) $x - y = 8$

Answer

$(9, 1)$

Solution

Adding the two equations we get $2x = 18$ so $x = 9$. Hence $9 + y = 10$ so $y = 1$. Hence the solution to the system is $(9, 1)$.

(b) (i) $6a + 5b = 7$, (ii) $3a + 5b = 1$

Answer

$(2, -1)$

Solution

Subtracting the second equation from the first we have $3a = 6$ so $a = 2$. Hence $6(2) + 5b = 7$ so $5b = -5$ so $b = -1$. This gives $(2, -1)$ as a solution.

Problem 6.4 Solve the following equations using elimination.

(a) (i) $4x + y = 11$, (ii) $5x - 2y = 4$.

Answer

$(2, 3)$

Solution

Multiplying the first equation by 2 gives $8x + 2y = 22$. Adding this to the second gives us $13x = 26$ so $x = 2$ after dividing by 13. Therefore $4(2) + y = 11$ so $y = 11 - 8 = 3$, giving $(2, 3)$ as our solution.

(b) (i) $\frac{1}{2}x - \frac{2}{3}y = \frac{7}{3}$, (ii) $\frac{3}{2}x + 2y = -25$

Answer

$(-6, -8)$

Solution

Multiplying the first equation by 3 gives $\frac{3}{2}x - 2y = 7$. Subtracting this from the second equation we get $4y = -32$ so $y = -8$. Plugging this back into the equation $\frac{3}{2}x - 2y = 7$

we have

$$\frac{3}{2}x - 2(-8) = 7 \text{ or } \frac{3}{2}x + 16 = 7 \text{ so } \frac{3}{2}x = -9 \text{ and thus } x = \frac{2}{3} \times -9 = -6.$$

Hence $(6, -8)$ is our solution.

Problem 6.5 Graph the solution sets to each of the inequalities below.

(a) (i) $y \geq x - 3$, (ii) $y > -x + 1$.

Answer

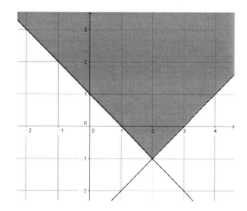

Solution

We know the lines respectively have slope 1 and -1 with y-intercepts $-$ and 1 which gives the graphs shown above. Note the lines intersect when $x - 3 = -x + 1$ or $2x = 4$ so $x = 2$. The y-value of this intersection is $x - 3 = -1$. We want the region above both lines which gives the shaded region above.

(b) (i) $y \leq 2x + 3$, (ii) $y > -3x - 2$.

Answer

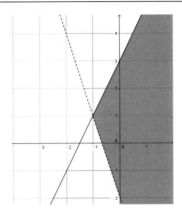

Solution

We know the lines respectively have slope 2 and -3 with y-intercepts 3 and -2 which gives the graphs shown above. Note the lines intersect when $2x + 3 = -3x - 2$ or $5x = -5$ so $x = -1$. The y-value of this intersection is $2(-1) + 3 = 1$. We want the region below the first line and above the second, resulting in the shaded region shown.

Problem 6.6 Graph the solution sets for the following.

(a) (i) $y \geq |x| - 3$, (ii) $x - y \geq -1$.

Answer

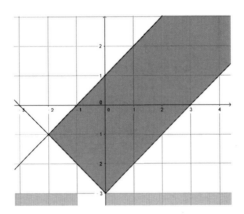

Solution

For (ii), we can rearrange to get $x + 1 \geq y$ or $y \leq x + 1$, so we want the region below the line $y = x + 1$ (which has slope 1 and y-intercept 1). Note that $|x| - 3$ is equal to $x - 3$

when $x \geq 0$ and $-x-3$ when $x < 0$. Hence we want the region above the lines $y = x - 3$ and $y = -x - 3$ (lines with slopes ± 1 and y-intercept 1). Combining gives the shaded region above.

(b) (i) $y \leq 2x - 1$, (ii) $4x - 2y > -2$

Answer

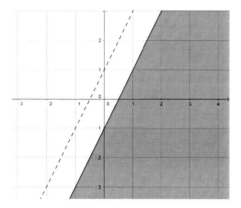

Solution

First note we want the region below the line $y = 2x - 1$ (which has slope 2 and y-intercept -1). Rewriting the second inequality we have $-2y > -4x - 2$ or $y < 2x + 1$ (remember to flip the sign when dividing by a negative number). Hence we want the region below the line $y = 2x + 1$ (with slope 2 and y-intercept 1). However, as the lines $y = 2x - 1$ and $y = 2x + 1$ are parallel, we just want the region below $y = 2x - 1$, which is seen shaded above.

Problem 6.7 Solve the system if (i) $x + y + z = 10$, (ii) $x = 4y$, and (iii) $z = -y$.

Answer

$(10, 5/2, -5/2)$

Solution

Substituting $x = 4y$ and $z = -y$ we get an equation only in terms of y: $4y + y - y = 10$. Hence $y = 10/4 = 5/2$. Thus, $x = 4 \times (5/2) = 10$ and $z = -5/2$.

Problem 6.8 Solve the following systems of equations with three variables and three unknowns.

(a) (i) $x+y+z=3$, (ii) $x+2y+3z=4$, and (iii) $x+2y+4z=5$.

Answer

$(3,-1,1)$

Solution

Subtracting the first equation from the second and third equations, we get the two equations $y+2z=1$ and $y+3z=2$. Subtracting these two equations gives $z=1$. Thus, $y+2\times1=1$ so $y=-1$. Lastly, $x=3-(-1)-1=3$.

(b) (i) $6x+2y-z=3$, (ii) $4x+3y+z=7$, and (iii) $z=x+3$.

Answer

$(2,-2,5)$

Solution

Substituting $z=x+3$ into the first two equations we have $5x+2y=6$ and $5x+3y=4$. Subtracting we get $-y=2$ so $y=-2$. Hence $5x+2\times(-2)=6$ so $5x=10$ and $x=2$. Finally, $z=2+3=5$.

Problem 6.9 Consider the equations: (i) $x+y-z=4$, (ii) $2x+2y-2z=4$, and (iii) $x+y-2z=4$. Which pairs of the pairs, (i) & (ii), (i) & (iii), (ii) & (iii), have solutions? Does the system of all three equations have a solution?

Answer

(i) & (iii) and (ii) & (iii) have solutions, (i) & (ii) does not. The full system does not have a solution.

Solution

Note dividing (ii) by 2 we get $x+y-z=2$. Hence (i) says $x+y-z=4$ but (ii) says $x+y-z=2$ which is impossible, so (i) & (ii) does not have any solutions.

For (i) & (iii) note subtracting (i) from (iii) gives $(x-x)+(y-y)+(-2z-(-z))=(4-4)$ so $-z=0$ and hence $z=0$. Substituting this in we have $x+y=4$. Hence any numbers x,y,z such that $x+y=4$ and $z=0$ will be a solution to (i) & (iii). (For example $(2,2,0)$ is one solution.)

For (ii) & (iii) note subtracting the simplified (ii) from (iii) gives $(x-x)+(y-y)+(-2z-(-z))=(4-2)$ so $-z=2$ and hence $z=-2$. Substituting this in we have $x+y=0$. Hence any numbers x,y,z such that $x+y=0$ and $z=-2$ will be a solution to (ii) & (iii). (For example $(-2,2,-2)$ is one solution.)

Note we have already seen it is impossible to make (i) & (ii) true at the same time, so the full system of equations does not have a solution.

Problem 6.10 Solve the following systems using some of the techniques from today.

(a) (i) $x^2-y^2=4$, (ii) $x^2+y^2=4$.

Answer

$(2,0),(-2,0)$

Solution

Adding the two equations we have $2x^2=8$ so $x^2=4$. We see that both $x=2$ and $x=-2$ work. Substituting these in we see that $4-y^2=4$ so $-y^2=0$ and hence $y=0$ in either case. Hence our solutions are $(2,0)$ and $(-2,0)$.

(b) (i) $|2x|+y=x+y+2$, and (ii) $y=6x-4$.

Answer

$(2,8)$ and $(-2/3,-8)$

Solution

Simplifying the first equation we have $|2x|=x+2$. If $x>0$ we have $2x=x+2$ so $x=2$ while if $x<0$ we have $-2x=x+2$ so $x=-2/3$. Using (ii), if $x=2$ then $y=6(2)-4=8$ and if $x=-2/3$ then $y=6(-2/3)-4=-8$. Hence our solutions are $(2,8)$ and $(-2/3,-8)$.

7 Solutions to Chapter 7 Examples

Problem 7.1 Systems of Equations Review

(a) Solve (i) $x - y + z = 5$, (ii) $y - z = 3$, (iii) $x = 4$.

Answer

No solutions exist.

Solution

We know $x = 4$ so substituting this into the first equation we have $4 - y + z = 5$ or $-y + z = 1$ or $y - z = -1$. However, the second equation says $y - z = 3$, so this is impossible. Hence no solutions exist.

(b) Solve (i) $3x - 4y + z = 8$, (ii) $2x + 4y + z = -1$, (iii) $x - 2y + 3z = 6$.

Answer

$(1, -1, 1)$

Solution

If we double the third equation we get $2x - 4y + 6z = 12$. Then adding the second equation to the first and third we have

$$(2x + 3x) + (4y - 4y) + (z + z) = (-1 + 8) \text{ or } 5x + 2z = 7$$

and similarly

$$(2x + 2x) + (4y - 4y) + (z + 6z) = (-1 + 12) \text{ or } 4x + 7z = 11.$$

Multiplying these by 4 and 5 respectively gives $20x + 8z = 28$ and $20x + 35z = 55$ so subtracting we have $(20x - 20x) + (35z - 8z) = (55 - 28)$ so $27z = 27$ and hence $z = 1$. Substituting back in we get $5x + 2(1) = 7$ so $5x = 5$ and $x = 1$. Lastly we have $3(1) - 4y + (1) = 8$ so $-4y = 4$ and $y = -1$. This gives $(1, -1, 1)$ as our solution.

(c) Solve (i) $2|x| + y = 5$, (ii) $|x| - y = 1$.

Answer

$(2, 1), (-2, 1)$

Solution

Adding the two equations we have $(2|x| + |x|) + (y - y) = 5 + 1$ so $3|x| = 6$ or $|x| = 2$. Hence $x = 2$ or $x = -2$. In either case we have $2(2) + y = 5$ so $y = 1$. Hence our solutions are $(2, 1)$ and $(-2, 1)$.

Problem 7.2 Solve the following questions (i) using a system of equations, (ii) without using a system of equations. Compare the two methods.

(a) Suppose there are chicken, rabbits, and sheep on a farm. There are 70 heads in total and 220 feet. If there are the same number of rabbits and sheep, how many chickens, rabbits, and sheep are on the farm?

Answer

30 chickens, 20 rabbits, 20 sheep

Solution 1

There are an equal number of rabbits and sheep, so pair up one rabbit with one sheep. This pair has 2 heads and 8 legs. If there are no chickens, there are

$$70 \div 2 = 35$$

such pairs, for a total of

$$35 \times 8 = 280$$

feet. This is a total of

$$280 - 220 = 60$$

extra feet. Replacing a rabbit/sheep pair with 2 chickens reduces the number of feet by

$$8 - 4 = 4.$$

Hence there are

$$60 \div 4 = 15$$

pairs of chickens and thus

$$35 - 15 = 20$$

rabbit/sheep pairs. Thus there are 30 chickens, 20 rabbits, and 20 sheep on the farm.

Solution 2

Let x, y, z be the number of chickens, rabbits, and sheep on the farm. Hence $x + y + z = 70$ as there are 70 heads in total, and $2x + 4y + 4z = 220$ for the number of feet. We also know that $y = z$ as there are the same number of rabbits and sheep. Hence we know $x + y + y = 70$ so $x + 2y = 70$ and similarly $2x + 4y + 4y = 220$ so $2x + 8y = 220$. Doubling the first equation we have $2x + 4y = 140$ so subtracting this from the second we have $(2x - 2x) + (8y - 4y) = (220 - 140)$ and hence $4y = 80$ and thus $y = 20$. Therefore $z = 20$ and lastly $x + 20 + 20 = 70$ so $x = 30$.

(b) Some chickens and rabbits have a total of 100 feet. If each chicken was exchanged for a rabbit, and each rabbit was exchanged for a chicken, there would be a total of 86 feet. How many chickens are there? How many rabbits?

Answer

12 chickens, 19 rabbits

Solution 1

We know that there are 100 feet total. We first find out how many animals there are in total. Pretend that for every chicken on the farm, we pair it up with a new rabbit, and for every rabbit on the farm, we pair it up with a new chicken. Note that this adds a total of 86 feet. Hence there are a total of

$$100 + 86 = 186$$

feet. Each chicken and rabbit pair has a combined total of

$$2 + 4 = 6$$

feet, so there must be

$$186 \div 6 = 31$$

chicken and rabbit pairs. As every original animal is in exactly one pair, this means there are 31 animals on the farm.

If all 31 animals were chickens, there would be a total of

$$31 \times 2 = 62$$

feet, which is

$$100 - 62 = 38$$

less than the true amount. As each rabbit has

$$4 - 2 = 2$$

extra feet, if we change

$$38 \div 2 = 19$$

chickens to rabbits we will have the correct number of feet. Hence there are

$$31 - 19 = 12$$

chickens and 19 rabbits.

Solution 2

(Algebra) Let x be the number of chickens and y be the number of rabbits. There are 100 feet in total, so as each chicken has 2 feet and each rabbit has 4,

$$2 \times x + 4 \times y = 100.$$

We also know that swapping all the animals we have 86 feet, so

$$4 \times x + 2 \times y = 86.$$

This gives the system of equations

$$\begin{cases} 2x + 4y &= 100, \\ 4x + 2y &= 86. \end{cases}$$

Doubling the first equation gives us

$$4x + 8y = 200.$$

We can then subtract the second equation from this to get

$$6y = 114,$$

and dividing by 6 we have

$$y = \frac{114}{6} = 19.$$

Substituting into the first equation,

$$2x + 4 \times 19 = 100$$

so combining like terms we have

$$2x = 24.$$

Hence

$$x = \frac{24}{2} = 12$$

so there are 12 chickens and 19 rabbits.

Problem 7.3 Solve the following questions (i) using a system of equations, (ii) without using a system of equations. Compare the two methods.

(a) Sami and Rajan practice running together. If Sami starts to run after Rajan runs for 10 meters, then it will take Sami 5 seconds to catch up with Rajan. If Sami starts to run after Rajan runs for 2 seconds, then it will take Sami 4 seconds to catch up with Rajan. How fast can each person run?

Answer

Sami: 6 meters/sec, Rajan: 4 meters/sec

Solution 1

In the second situation, Sami and Rajan run the same distance if Sami runs for 4 seconds and Rajan runs for 6 seconds. Thus, the speed at which they run is in the opposite ratio

$$6 : 4 = 3 : 2.$$

This also tells us that if they run the same amount of time, the ratio of the distances Sami and Rajan run is also

$$3 : 2.$$

In the first scenario, Sami needs to run 10 extra meters and Rajan (both running for 5 seconds). Multiplying the above ratio by 10 we get

$$30 : 20$$

so Sami runs for 30 meters and Rajan runs 20 meters. Since this takes them each 5 seconds, we see that Sami runs at a rate of

$$30 \div 5 = 6$$

meters/sec and Rajan runs at a rate of

$$20 \div 5 = 4$$

meters/sec.

Solution 2

(Algebra) Let's assume the Sami's running speed is x meters/sec, and Rajan's running speed is y meters/sec. In the first situation, Sami runs a distance of $5 \times x$, while Rajan runs a distance of $10 + 5 \times y$, so

$$5 \times x = 10 + 5 \times y.$$

In the second situation, Sami runs for 4 seconds, while Rajan runs for 6 seconds, so

$$4 \times x = 6 \times y.$$

This gives us the system of equations

$$\begin{cases} 5x &= 5y + 10, \\ 4x &= 6y. \end{cases}$$

Dividing the first equation by 5 gives us

$$x = y + 2$$

so we can substitute this into the second equation to get

$$4 \times (y + 2) = 6y.$$

Distributing and combining like terms we have

$$8 = 2y$$

so we can solve for y and get

$$y = 4.$$

Substituting back into the second equation we have

$$4x = 6 \times 4$$

so we can solve for x:

$$x = 6.$$

Therefore, Sami runs at a speed of 6 meters/sec, and Rajan runs at a speed of 4 meters/sec.

(b) When two teams A and B work together, it takes 18 days to get a job completed. After team A works for 3 days, and team B works for 4 days, only $\frac{1}{5}$ of the job is done. How long does it take for team A alone to complete the job? For team B alone?

Answer

A: 45 days, B: 30 days

Solution 1

We know that if team A works for 3 days and team B works for 4 days they can finish $\frac{1}{5}$ of the job. If they work 6 times longer, team A works for 18 days and team B works for 24 days and they complete $\frac{6}{5}$ of a job. Combined with the fact that the two teams can complete the job in 18 days, we have that if team B works $24 - 18 = 6$ days it can complete $\frac{6}{5} - 1 = \frac{1}{5}$ of a job. Thus team B can complete

$$\frac{1}{5} \div 6 = \frac{1}{30}$$

of a job in one day. In 18 days team B completes

$$18 \times \frac{1}{30} = \frac{3}{5}$$

of a job, so team A must complete $\frac{2}{5}$ of a job in 18 days, and

$$\frac{2}{5} \div 18 = \frac{1}{45}$$

of a job in one day. We then know it takes team A 45 days and team B 30 days to complete the job if they work alone.

Solution 2

Let x denote the amount of work team A can complete in one day and y the amount of work team B can complete in one day. Working together they can complete the job in 18 days so

$$18 \times x + 18 \times y = 1.$$

If team A works for 3 and team B works for 4 days, $\frac{1}{5}$ of the job is completed, so

$$3 \times x + 4 \times y = \frac{1}{5}.$$

This gives us the system of equations

$$\begin{cases} 18x + 18y & = & 1, \\ 3x + 4y & = & \frac{1}{5}. \end{cases}$$

Multiplying the second equation by 6, we get

$$18x + 24y = \frac{6}{5}.$$

Subtracting the first equation from this equation we have

$$6y = \frac{1}{5}.$$

so

$$y = \frac{1}{30}.$$

Substituting back into the first equation,

$$18x + 18 \times \frac{1}{30} = 1,$$

so we can solve for x to get

$$x = \frac{1}{45}.$$

Therefore it takes team A 45 days working alone and team B 30 days working alone to complete the job.

Problem 7.4 George starts to build his investment portfolio. He has $10,000 divided between stocks, bonds, and savings. He earns a 10% return on stocks, 6% on bonds, and 1% on savings. His total return was $520. If his total investment in stocks equaled his total investment in bonds, how much did George keep in savings?

Answer

$4000

Solution

Let x, y, z stand for how much George invests in stocks, bonds, and savings respectively. We then have $x + y + z = 10000$. We also know that $x = y$ because George invests equally in stocks and bonds. His total return is 520 dollars, so $10\% \times x + 6\% \times y + 1\% \times z = 520$ and hence $0.1x + 0.06y + 0.01z = 520$. To get rid of all the decimals, multiply this third equation by 100 to get $10x + 6y + z = 52000$. Subtracting the first equation from this, we have $9x + 5y = 42000$. As we also know $x = y$ we have $9x + 5x = 42000$ so $14x = 42000$ and $x = 3000$. Hence $y = 3000$ as well. Lastly we get $3000 + 3000 + z = 10000$ so $z = 4000$, the amount George kept in savings.

Problem 7.5 Wilson has a total of 60 chickens and rabbits on a farm. One day he buys some number of chickens so that he has the same number of chickens as rabbits. The next day he buys the same number of chickens again. At this point he has 40 more rabbit legs than chicken legs. How many rabbits does Wilson have?

Answer

40 rabbits

Solution

Let x be the number of chickens and y the number of rabbits Wilson initially has on the farm. Then we know $x + y = 60$. To get the same number of chickens as rabbits he must buy $y - x$ chickens. This means he also buys $y - x$ chickens the next day as well. Hence he ends up with $x + y - x + y - x = 2y - x$ chickens and y rabbits. At this point there are 40 more rabbit legs than chicken legs, so $2 \cdot (2y - x) + 40 = 4 \cdot y$. The second equation can be simplified to $4y - 2x + 40 = 4y$ so cancelling the $4y$'s we get $-2x + 40 = 0$ so $x = 20$. Hence $y = 60 - 20 = 40$ and Wilson has 40 rabbits.

Problem 7.6 There are 50 bugs in total, made up of spiders (8 legs, no wings), houseflies (6 legs, 1 pair of wings), and dragonflies (6 legs, 2 pair of wings). There is one housefly for every dragonfly leg. If every spider is exchanged with a dragonfly, there is one housefly for every pair of dragonfly wings. How many dragonflies are there?

Answer

5

Solution

Suppose there are x spiders, y houseflies, and z dragonflies. There are 50 bugs in total, so $x + y + z = 50$. Dragonflies have 6 legs, so $y = 6z$. Similarly dragonflies have 2 pairs of wings, so $y = 2x$ (after exchanging spiders and dragonflies). We know $y = 6z$ and since $2x = y = 6z$ we also have $x = 3z$. Plugging these into the first equation we get $(3z) + (6z) + z = 50$ so $10z = 50$ and $z = 5$, the number of dragonflies.

Problem 7.7 Alice, Bob, and Eve are washing cars. In total they wash 300 cars. Alice washes a third as many cars as Bob and Eve combined. Eve washes half as many cars as Alice and Bob combined. How many cars does Bob wash?

Answer

125 cars

Solution

Let x, y, z be the number of cars Alice, Bob, Eve wash. We know $x + y + z = 300$, $3x = y + z$, and $2z = x + y$ from the given information. Adding the first and third equations we get $x + y + 3z = 300 + x + y$ so canceling we have $3z = 300$ so $z = 100$. Putting this back into the second and third equations we have $3x = y + 100$ and $x + y = 200$. The second says $x = 200 - y$ and thus (substituting) $3(200 - y) = y + 100$ or $600 - 3y = y + 100$. Combining like terms we have $500 = 4y$ so dividing we have $y = 125$.

Problem 7.8 Troy and Max work together to paint a fence. If Troy works alone for 5 days and Max works for 3 days, they can finish the entire fence. If Max works for 1 day and then they work together for 1 day, they can finish $1/4$ of the fence. How long does it take Max to paint the fence alone?

Answer

28 days

Solution

Let x be the amount of the fence Troy can paint in one day and y the amount Max can paint in one day. The given information says $5x + 3y = 1$ and $1x + 2y = 1/4$. Solving the second for x we have $x = 1/4 - 2y$ so if we plug this into the first equation we get $5(1/4 - 2y) + 3y = 1$ so distributing and combining like terms we have $-7y = -1/4$ so $y = 1/28$. Thus Max can paint $1/28$ of the fence in one day so it takes 28 days to paint the entire fence.

Problem 7.9 Bill, Claire, and Drew went on a road trip over the weekend. In total they drove 500 miles over 10 hours. Bill drove at an average rate of 40 mph, Claire an average of 55 mph, and Drew an average of 65 mph. If Bill and Drew drove the same distance, how long many hours did each drive?

Answer

Bill: 130/23, Claire: 20/23, Drew: 80/23.

Solution

Let x, y, z denote the number of hours Bill, Claire, Drew drove. Hene from the information given we know $x + y + z = 10$, $40x + 55y + 65z = 500$, and $40x = 65z$. Multiplying the first equation by 55 we get $55x + 55y + 55z = 550$ and if we subtract the second equation from this we get $15x - 10z = 50$. Multiplying this by 8 and the third equation by 3 we get $120x - 80z = 400$ and $120x = 195z$ so subtracting we get $-80z = 400 - 195z$ so $115z = 400$ so $z = \dfrac{400}{115} = \dfrac{80}{23}$ (Drew's number of hours). We can then substitute this to solve for $y = \dfrac{20}{23}$ (Claire's number of hours). Finally we can use y, z to find $x = \dfrac{130}{23}$ (Bill's number of hours).

Problem 7.10 Farmer Billy has a farm where he raises chickens and rabbits. One Sunday Billy counted 500 legs on the farm. On Monday he traded all his rabbits for chickens (1 rabbit for 1 chicken). On Tuesday, he traded all his original chickens (not the new ones) so that he received 3 rabbits for every 2 chickens. On Wednesday Billy remarked that there were still 500 legs on the farm. How many chickens and how many rabbits did Billy start the week with?

Answer

50 chickens and 100 rabbits

Solution

Let x be the number of chickens and y the number of rabbits. Hence counting legs on Sunday we have $2x + 4y = 500$. On Monday he ends up with x old chickens and y new chickens. On Tuesday he ends up with y chickens and $1.5x$ rabbits, so counting legs again on Wednesday we have $2y + 4(1.5x) = 500$ or equivalently $6x + 2y = 500$. Doubling this equation and subtracting it from the first we have $-10x = -500$ so $x = 50$. We can then solve for $y = 100$.

8   Solutions to Chapter 8 Examples

Problem 8.1 Simplify the following expressions. You may assume any variables denote positive numbers.

(a) $(5x^2)^3$

Answer

$125x^6$

Solution

We have that $(5x^2)^3 = 5^3 \times (x^2)^3 = 125x^{2 \times 3} = 125x^6$.

(b) $\sqrt{169}$

Answer

13

Solution

Note that $13^2 = 369$ so $\sqrt{369} = 13$.

(c) $\sqrt{16^3}$

Answer

64

Solution

Note that $16^3 = 4096$ and $64^2 = 4096$ so $\sqrt{4096} = 64$.

Alternatively, $\sqrt{16^3} = \sqrt{16}^3 = 4^3 = 64$.

(d) $\sqrt{9x^4}$

Answer

$3x^2$

Solution

Note that $(3x^2)^2 = 9x^4$ so $\sqrt{9x^4} = 3x^2$.

Problem 8.2 Write the following in simplest radical form.

(a) $\sqrt{32}$

Answer

$4\sqrt{2}$

Solution

Note $32 = 16 \times 2$ so $\sqrt{32} = \sqrt{16 \times 2} = \sqrt{16} \times \sqrt{2} = 4\sqrt{2}$.

(b) $3\sqrt{63}$

Answer

$9\sqrt{7}$

Solution

Note $\sqrt{63} = \sqrt{9 \times 7} = \sqrt{9} \times \sqrt{7} = 3\sqrt{7}$. Hence $3\sqrt{63} = 3 \times 3\sqrt{7} = 9\sqrt{7}$.

(c) $\sqrt{72}$

Answer

$6\sqrt{2}$

Solution

We have $\sqrt{72} = \sqrt{36 \times 2} = \sqrt{36} \times \sqrt{2} = 6\sqrt{2}$.

Problem 8.3 Write the following expressions in simplest radical form. You may assume that any of the variables used in the problems are positive.

(a) $\sqrt{20x^5}$

Answer

$2x^2\sqrt{5x}$

Solution

We have $20x^5 = 4x^4 \times 5x$ so $\sqrt{20x^5} = \sqrt{4x^4 \times 5x} = \sqrt{4x^4} \times \sqrt{5x}$. Since $2x^2 \times 2x^2 = 4x^4$ we have $\sqrt{4x^4} = 2x^2$ giving a final answer of $2x^2\sqrt{5x}$.

(b) $4\sqrt{243x^3y^5}$

Answer

$36xy^2\sqrt{3xy}$

Solution

We have $\sqrt{243x^3y^5} = \sqrt{81x^2y^4 \times 3xy} = \sqrt{81x^2y^4} \times \sqrt{3xy} = 9xy^2\sqrt{3xy}$. Hence

$$4\sqrt{243x^3y^5} = 4 \times 9xy^2\sqrt{3xy} = 36xy^2\sqrt{3xy}.$$

Problem 8.4 Expand and simplify the following expressions.

(a) $3x^2(x+y+2)$.

Answer

$3x^3 + 3x^2y + 6x^2$

Solution

Distributing we have $3x^2 \times x + 3x^2 \times y + 3x^2 \times 2 = 3x^3 + 3x^2y + 6x^2$.

(b) $2xy^2(x+3y) + 2x^2(3x-y+5)$.

Answer

$6x^3 + 2x^2y^2 - 2x^2y + 10x^2 + 6xy^3$

Solution

Distributing we have $2xy^2(x+3y) = 2xy^2 \times x + 2xy^2 \times 3y = 2x^2y^2 + 6xy^3$. Similarly

$2x^2(3x - y + 5) = 6x^3 - 2x^2y + 10x^2$. Combining we get a final answer of $6x^3 + 2x^2y^2 - 2x^2y + 10x^2 + 6xy^3$.

Problem 8.5 Adding and Subtracting Radicals

(a) $2\sqrt{8} - \sqrt{2} + 3\sqrt{18}$

Answer

$12\sqrt{2}$

Solution

First note $2\sqrt{8} = 2\sqrt{4}\sqrt{2} = 4\sqrt{2}$. Similarly, $3\sqrt{18} = 3\sqrt{9}\sqrt{2} = 9\sqrt{2}$. Hence our expression becomes
$$4\sqrt{2} - \sqrt{2} + 9\sqrt{2} = 12\sqrt{2}.$$

(b) $4\sqrt{24} + \sqrt{27} - \sqrt{54}$

Answer

$5\sqrt{6} + 3\sqrt{3}$

Solution

We have $4\sqrt{24} = 4\sqrt{4}\sqrt{6} = 8\sqrt{6}$. Similarly $\sqrt{27} = \sqrt{9}\sqrt{3} = 3\sqrt{3}$ and $\sqrt{54} = \sqrt{9}\sqrt{6} = 3\sqrt{6}$. Hence we have $8\sqrt{6} + 3\sqrt{3} - 3\sqrt{6}$ which after combining identical radicals gives $5\sqrt{6} + 3\sqrt{3}$.

Problem 8.6 Multiplying Radicals. Assume that any variables denote positive numbers.

(a) $\sqrt{2} \times \sqrt{8}$

Answer

4

Solution

We have $\sqrt{2} \times \sqrt{8} = \sqrt{2 \times 8} = \sqrt{16} = 4$.

(b) $2\sqrt{15}(r+\sqrt{5})$

Answer

$2r\sqrt{15}+10\sqrt{3}$

Solution

Distributing we have $2\sqrt{15}(r+\sqrt{5})=2r\sqrt{15}+2\sqrt{15\times5}$. Since

$$2\sqrt{15\times5}=2\sqrt{3\times5\times5}=2\times5\sqrt{3}=10\sqrt{3}$$

we have a final answer of $2r\sqrt{15}+10\sqrt{3}$.

(c) $\sqrt{3r}(\sqrt{r^3}+\sqrt{3})$

Answer

$r^2\sqrt{3}+3\sqrt{r}$

Solution

Distributing and combining into single radicals gives $\sqrt{3r^4}+\sqrt{9r}=r^2\sqrt{3}+3\sqrt{r}$.

Problem 8.7 Exponents in Fractions

(a) $\dfrac{125x^3y^2}{5xy^2}$

Answer

$25x^2$

Solution

We have

$$\frac{125}{5}\times\frac{x^3}{x}\times\frac{y^2}{y^2}=5\times x^2\times1=5x^2.$$

(b) $\dfrac{80x^4y^2z^5}{120xy^4z^3}$

Answer

$$\frac{2x^3z^2}{3y^2}$$

Solution

We have

$$\frac{80}{120} \times \frac{x^4}{x} \times \frac{y^2}{y^4} \times \frac{z^5}{z^3} = \frac{2}{3} \times x^3 \times \frac{1}{y^2} \times z^2 = \frac{2x^3z^2}{3y^2}.$$

Problem 8.8 Simplifying Radicals in Fractions

(a) $\dfrac{\sqrt{18}-27}{3}$

Answer

$\sqrt{2}-9$

Solution

First we have $\sqrt{18} = \sqrt{9}\sqrt{2} = 3\sqrt{2}$. Hence after dividing each term by 3 we have $\sqrt{2}-9$ as our final answer.

(b) $\dfrac{\sqrt{6}-x\sqrt{3}}{\sqrt{3}}$

Answer

$\sqrt{2}-x$

Solution

Dividing we have

$$\frac{\sqrt{6}-x\sqrt{3}}{\sqrt{3}} = \frac{\sqrt{6}}{\sqrt{3}} - \frac{x\sqrt{3}}{\sqrt{3}} = \sqrt{\frac{6}{3}} - x\sqrt{1} = \sqrt{2}-x.$$

(c) $\dfrac{x^2 - \sqrt{y}}{2\sqrt{7y}}$. Here you may assume $y > 0$.

Answer

$$\frac{x^2\sqrt{7y} - y\sqrt{7}}{14y}$$

Solution

We first multiply the numerator and denominator by $\sqrt{7y}$ to get

$$\frac{x^2\sqrt{7y} - \sqrt{y}\sqrt{7y}}{2 \times 7y} = \frac{x^2\sqrt{7y} - y\sqrt{7}}{14y}$$

as our final answer.

Problem 8.9 Higher Power Roots

(a) Simplify $\sqrt[3]{108}$

Answer

$3\sqrt[3]{4}$

Solution

Noting that $3^3 = 27$ and $108 = 27 \times 4$ we have that $\sqrt[3]{108} = \sqrt[3]{27}\sqrt[3]{4} = 3\sqrt[3]{4}$.

(b) $\dfrac{\sqrt[3]{15}}{\sqrt[3]{64}}$

Answer

$\dfrac{\sqrt[3]{15}}{4}$

Solution

We have that $4^3 = 64$ so in fact $\sqrt[3]{64} = 4$. As there are no perfect cubes that are factors of 15, we can simplify the expression to $\dfrac{\sqrt[3]{15}}{4}$.

(c) $\dfrac{\sqrt[3]{10}}{\sqrt[3]{4}}$

Answer

$$\frac{\sqrt[3]{20}}{2}$$

Solution

Note $2^3 = 8$ so if we multiply the numerator and denominator by $\sqrt[3]{2}$ we can simplify the denominator

$$\frac{\sqrt[3]{10}}{\sqrt[3]{4}} = \frac{\sqrt[3]{10}}{\sqrt[3]{4}} \times \frac{\sqrt[3]{2}}{\sqrt[3]{2}} = \frac{\sqrt[3]{20}}{\sqrt[3]{8}} = \frac{\sqrt[3]{20}}{2}.$$

As 20 does not contain any factors that are perfect cubes, this is our final answer.

Problem 8.10 Higher Power Radicals with Expressions. Assume any variables denote positive numbers.

(a) $2\sqrt[4]{243x^5y^{12}z^3}$

Answer

$6xy^3 \sqrt[4]{3xz^3}$

Solution

We have $243x^5y^{12}z^3 = 81x^4y^{12} \times 3xz^3 = (3xy^3)^4 \times 3xz^3$. Hence

$$2\sqrt[4]{243x^5y^{12}z^3} = 2\sqrt[4]{(3xy^3)^4 \times 3xz^3} = 2(3xy^3)\sqrt[4]{3xz^3} = 6xy^3\sqrt[4]{3xz^3}.$$

(b) $\dfrac{4\sqrt{4m^2n^4}}{\sqrt[4]{64m^4n^2}}$

Answer

$2n\sqrt[4]{4n^2}$

Solution

We first note that $\sqrt{4m^2n^4} = 2mn^2$ so the numerator is just $4\sqrt{4m^2n^4} = 8mn^2$. For the denominator, $\sqrt[4]{64m^4n^2} = \sqrt[4]{16m^4}\sqrt[4]{4n^2} = 2m\sqrt[4]{4n^2}$. Simplifying we have

$$\frac{8mn^2}{2m\sqrt[4]{4n^2}} = \frac{4n^2}{\sqrt[4]{4n^2}} \times \frac{\sqrt[4]{4n^2}}{\sqrt[4]{4n^2}} = \frac{4n^2\sqrt[4]{4n^2}}{\sqrt[4]{16n^4}} = \frac{4n^2\sqrt[4]{4n^2}}{2n} = 2n\sqrt[4]{4n^2}.$$

9 Solutions to Chapter 9 Examples

Problem 9.1 Review

(a) Write $\sqrt{756}$ in simplest radical form.

Answer

$6\sqrt{21}$

Solution

We have that $\sqrt{756} = \sqrt{36}\sqrt{21} = 6\sqrt{21}$ when written in simplest radical form.

(b) Write $\sqrt[3]{756}$ in simplest radical form.

Answer

$3\sqrt[3]{28}$

Solution

We have that $\sqrt[3]{756} = \sqrt[3]{27}\sqrt[3]{28} = 3\sqrt[3]{28}$ when in simplest radical form.

(c) Simplify $x^2(x+y+4)$

Answer

$x^3 + x^2y + 4x^2$

Solution

Distributing we have $x^2(x+y+4) = x^3 + x^2y + 4x^2$. As none of these terms can be combined, this is our final answer.

(d) Simplify $x\sqrt{20}(x+y\sqrt{2}+\sqrt{5})$

Answer

$2x^2\sqrt{5} + 2xy\sqrt{10} + 10x$

Solution

First note $x\sqrt{20}=x\sqrt{4}\sqrt{5}=2x\sqrt{5}$. Hence when distributing we have $2x^2\sqrt{5}+2xy\sqrt{10}+10x$.

Problem 9.2 Expand the following.

(a) $(x+2)(x+4)$

Answer

x^2+6x+8

Solution

Using FOIL (multiply firsts, multiply outsides, multiply insides, multiply lasts) we have $x\times x+x\times4+2\times x+2\times4=x^2+4x+2x+8=x^2+6x+8$.

(b) $(2x+y)(x+5)$

Answer

$2x^2+xy+10x+5y$

Solution

Using FOIL we have $2x\times x+2x\times5+y\times x+y\times5=2x^2+10x+xy+5y$.

Problem 9.3 Expand the following expressions with radicals.

(a) $(2+\sqrt{2})(1-\sqrt{3})$

Answer

$2+\sqrt{2}-2\sqrt{3}-\sqrt{6}$

Solution

Using FOIL we have $2\times1+2\times(-\sqrt{3})+\sqrt{2}\times1+\sqrt{2}\times(-\sqrt{3})=2-2\sqrt{3}+\sqrt{2}-\sqrt{6}$.

(b) $(\sqrt{2}-\sqrt{3})(\sqrt{10}-1)$

Answer

$2\sqrt{5} - \sqrt{2} - \sqrt{30} + \sqrt{3}$

Solution

Using FOIL we have $\sqrt{2} \times \sqrt{10} + \sqrt{2} \times (-1) + (-\sqrt{3}) \times \sqrt{10} + (-\sqrt{3}) \times (-1) = \sqrt{2}\sqrt{2}\sqrt{5} - \sqrt{2} - \sqrt{30} + \sqrt{3} = 2\sqrt{5} - \sqrt{2} - \sqrt{30} + \sqrt{3}$.

Problem 9.4 Expand the following. Note that these are very useful formulas!

(a) $(a+b)^2$

Answer

$a^2 + 2ab + b^2$

Solution

Using FOIL we have $a \times a + a \times b + b \times a + b \times b = a^2 + 2ab + b^2$.

(b) $(a+b)(a-b)$

Answer

$a^2 - b^2$

Solution

Using FOIL we have $a \times a - a \times b + b \times a - b \times b = a^2 - b^2$.

Problem 9.5 Expand the following, simplifying where possible.

(a) $(x^2 + 2)^2$

Answer

$x^4 + 4x^2 + 4$

Solution

Practicing using the formula, if $a = x^2$ and $b = 2$ we want $(a+b)^2 = a^2 + 2ab + b^2 = (x^2)^2 + 2(x^2)(2) + (2)^2 = x^4 + 4x^2 + 4$.

(b) $(\sqrt{2}+\sqrt{5})^2$

Answer

$7+2\sqrt{10}$

Solution

Practicing using the formula, if $a=\sqrt{2}$ and $b=\sqrt{5}$ we want $(a+b)^2=a^2+2ab+b^2=(\sqrt{2})^2+2\sqrt{2}\sqrt{5}+(\sqrt{5})^2=2+2\sqrt{10}+5=7+2\sqrt{10}$.

(c) $(4+\sqrt{3})(4-\sqrt{3})$

Answer

13

Solution

Practicing the formula, if $a=4$ and $b=\sqrt{3}$ we want $(a+b)(a-b)=a^2-b^2=4^2-(\sqrt{3})^2=16-3=13$.

Problem 9.6 Simplify the following expressions.

(a) $\dfrac{11}{\sqrt{3}-5}$

Answer

$-\dfrac{5}{2}-\dfrac{\sqrt{3}}{2}$

Solution

Note multiplying by the conjugate gives $(\sqrt{3}-5)(\sqrt{3}+5)=(\sqrt{3})^2-5^2=-22$. Hence

$$\frac{11}{\sqrt{3}-5}=\frac{11}{\sqrt{3}-5}\times\frac{\sqrt{3}+5}{\sqrt{3}+5}=\frac{11\sqrt{3}+55}{-22}=\frac{\sqrt{3}+5}{-2}=-\frac{5}{2}-\frac{\sqrt{3}}{2}.$$

(b) $\dfrac{4\sqrt{21}}{\sqrt{3}+\sqrt{7}}$

Answer

$7\sqrt{3} - 3\sqrt{7}$

Solution

Note $\sqrt{21} = \sqrt{3}\sqrt{7}$. We multiply the numerator and denominator by the conjugate $\sqrt{3} - \sqrt{7}$ to get

$$\frac{4\sqrt{3}\sqrt{7}}{\sqrt{3}+\sqrt{7}} \times \frac{\sqrt{3}-\sqrt{7}}{\sqrt{3}-\sqrt{7}} = \frac{12\sqrt{7}-28\sqrt{3}}{(\sqrt{3})^2-(\sqrt{7})^2} = \frac{12\sqrt{7}+28\sqrt{3}}{(-4} = 7\sqrt{3}-3\sqrt{7}.$$

Problem 9.7 Simplify $\dfrac{3-\sqrt{y}}{3+\sqrt{y}}$

Answer

$$\frac{9+y-6\sqrt{y}}{9-y}$$

Solution

The conjugate of $3+\sqrt{y}$ is $3-\sqrt{y}$ so multiplying the numerator and denominator we get

$$\frac{3-\sqrt{y}}{3+\sqrt{y}} \times \frac{3-\sqrt{y}}{3-\sqrt{y}} = \frac{(3-\sqrt{y})^2}{(3+\sqrt{y})(3-\sqrt{y})} = \frac{9-6\sqrt{y}+y}{9-y}.$$

Problem 9.8 Solve the following equations.

(a) $2x\sqrt{3} = \sqrt{6}$.

Answer

$x = \dfrac{\sqrt{2}}{2}$

Solution

Isolating x by dividing both sides by $2\sqrt{3}$ gives

$$x = \frac{\sqrt{6}}{2\sqrt{3}} = \frac{\sqrt{6/3}}{2} = \frac{\sqrt{2}}{2}.$$

(b) $x + x\sqrt{2} = 4$.

Answer

$4\sqrt{2} - 4$

Solution

Our equation is the same as $x(1 + \sqrt{2}) = 4$ so

$$x = \frac{4}{1 + \sqrt{2}} = \frac{4}{1 + \sqrt{2}} \times \frac{1 - \sqrt{2}}{1 - \sqrt{2}} = \frac{4 - 4\sqrt{2}}{1 - 2} = 4\sqrt{2} - 4.$$

Problem 9.9 Solve the following equations by squaring both sides.

(a) $3\sqrt{7} = \sqrt{-y}$.

Answer

$y = -63$

Solution

Squaring both sides we have $-y = (3\sqrt{7})^2 = 9 \times 7 = 63$ so $y = -63$.

(b) $4 + \sqrt{x + 4} = 0$

Answer

No solutions

Solution

Isolating the square root we have $\sqrt{x + 4} = -4$. However, at this point note that $\sqrt{x + 4}$ is non-negative by definition, so this equation is impossible and there are no solutions.

(c) The square root of the sum of a number and 7 is 8. What is the number?

Answer

57

Solution

Let x be the number. Hence $\sqrt{x+7} = 8$. Squaring both sides gives $x + 7 = 64$ so $x = 64 - 7 = 57$.

Problem 9.10 Expand the following.

(a) $(x+y+z)^2$

Answer

$x^2 + y^2 + z^2 + 2xy + 2xz + 2yz$.

Solution

Distributing we have

$$(x+y+z)(x+y+z) = x(x+y+z) + y(x+y+z) + z(x+y+z)$$
$$= x^2 + xy + xz + xy + y^2 + yz + xz + yz + z^2$$
$$= x^2 + y^2 + z^2 + 2xy + 2xz + 2yz.$$

(b) $(\sqrt{2} + \sqrt{3} - 1)^2$

Answer

$6 + 2\sqrt{6} - 2\sqrt{2} - 2\sqrt{3}$

Solution

Let $x = \sqrt{2}, y = \sqrt{3}, z = -1$, we apply the formula

$$(x+y+z)^2 = x^2 + y^2 + z^2 + 2xy + 2xz + 2yz.$$

Thus

$$(\sqrt{2} + \sqrt{3} - 1)^2 = (\sqrt{2})^2 + (\sqrt{3})^2 + (-1)^2 + 2\sqrt{2}\sqrt{3} + 2\sqrt{2}(-1) + 2\sqrt{3}(-1)$$
$$= 6 + 2\sqrt{6} - 2\sqrt{2} - 2\sqrt{3}.$$